Getting Started with Intel Galileo

Matt Richardson

Getting Started with Intel Galileo

by Matt Richardson

Printed in the United States of America.

Published by Maker Media, Inc., 1005 Gravenstein Highway North, Sebastopol, CA 95472.

Maker Media books may be purchased for educational, business, or sales promotional use. Online editions are also available for most titles (*http://my.safaribooksonline.com*). For more information, contact O'Reilly Media's corporate/institutional sales department: 800-998-9938 or *corporate@oreilly.com*.

Editor: Brian Jepson
Production Editor: Melanie Yarbrough
Proofreader: Gillian McGarvey
Cover Designer: Juliann Brown
Interior Designer: David Futato
Illustrator: Rebecca Demarest
Photographer: Matt Richardson
Cover Photographer: Jeffrey Braverman

March 2014: First Edition

Revision History for the First Edition:

2014-03-11: First release

See *http://oreilly.com/catalog/errata.csp?isbn=9781457183089* for release details.

ISBN: 978-1-457-18308-9

[LSI]

Contents

Preface vii

1/Introduction to Galileo 1

What Is Galileo? 2

Inputs and Outputs 2

Code 4

Communication 4

What Makes Galileo Different? 5

Sketching in Hardware 8

2/First Steps 9

Tour of the Board 10

Helpful Tools and Components 13

Writing Programs to Control Your Galileo 16

Getting Familiar with the Development Environment 17

Connecting the Board 19

Uploading Code 21

Taking It Further 23

3/Outputs 25

Back to Blinking: Digital Output 26

Setup and Loop 26

Variables 27

Pin Numbers 28

Circuits and the Flow of Electricity 34

pinMode() 37

digitalWrite() 38

delay() 39

Code and Syntax Notes 39

Going Further with Digital Output 41

Analog Output 42

analogWrite() 43

Code and Syntax Notes 47

Other Outputs. 50
Serial Data Output. 50
Controlling A/C Appliances with Relays. 54
Controlling Servos. 55
Looking at Linux. 60
Connecting via Telnet. 61
Working with Pins. 62
Taking It Further. 65

4/Inputs. 67
Switches: Digital Input. 68
digitalRead(). 73
Code and Syntax Notes. 74
Analog Input. 75
Potentiometers. 76
analogRead(). 80
Code and Syntax Notes. 81
Variable Resistors. 82
Code and Syntax Notes. 87
Going Further. 88

5/Going Further with Code. 91
Data Types. 92
int. 92
float. 92
long. 93
boolean. 93
char. 94
String Object. 94
millis(). 95
Other Loops. 95
while. 95
do... while. 95
for Loops. 96
More Serial. 99
Serial.available() and Serial.read(). 99
Taking It Further. 100

6/Getting Online. 103
Connecting and Testing an Ethernet Connection. 104

Connecting and Testing with a WiFi Connection. 105
Connecting Using Linux Commands. 107
system(). 108
Getting Galileo's IP Address Using system(). 108
Connecting to Servers. 110
How Many Days Until MAKE Comes Out?. 110
Serving a Web Page. 123
Serving a Web Page with Python. 126
Taking It Further. 129

A/ Arduino Code Reference. **131**

B/ Breadboard Basics. **147**

C/ Resistor Reference. **153**

D/ Creating a MicroSD Image. **157**

E/ Setting Up Galileo on Windows. **161**

F/ Setting Up Galileo on Linux. **167**

G/ Setting Up Galileo on Mac OS X. **171**

H/ Connecting to Galileo via Serial. **175**

Preface

Intel Galileo is a hardware development board that lets you write code and create electronic circuits to build your own projects. It's capable of acting as the brain in a robot, controlling haunted house special effects, uploading sensor data to the Internet, and much more.

The board doesn't do very much on its own, so it's up to you connect the right hardware and write the code to tell it what you want it to do. In that sense, Galileo is like a painter's canvas. It doesn't become anything remarkable until you start to work with it.

Luckily, since Galileo is Arduino-compatible, you have a vast amount of resources from the world of Arduino available to you. These include code examples, libraries that help you do complex things, expansion shields that make it easy to connect circuits, and a simple development workflow—which means you spend more time being creative and less time figuring out how to get things to work. Not only that, but you also have access to the enormous community of Arduino users if you run into trouble.

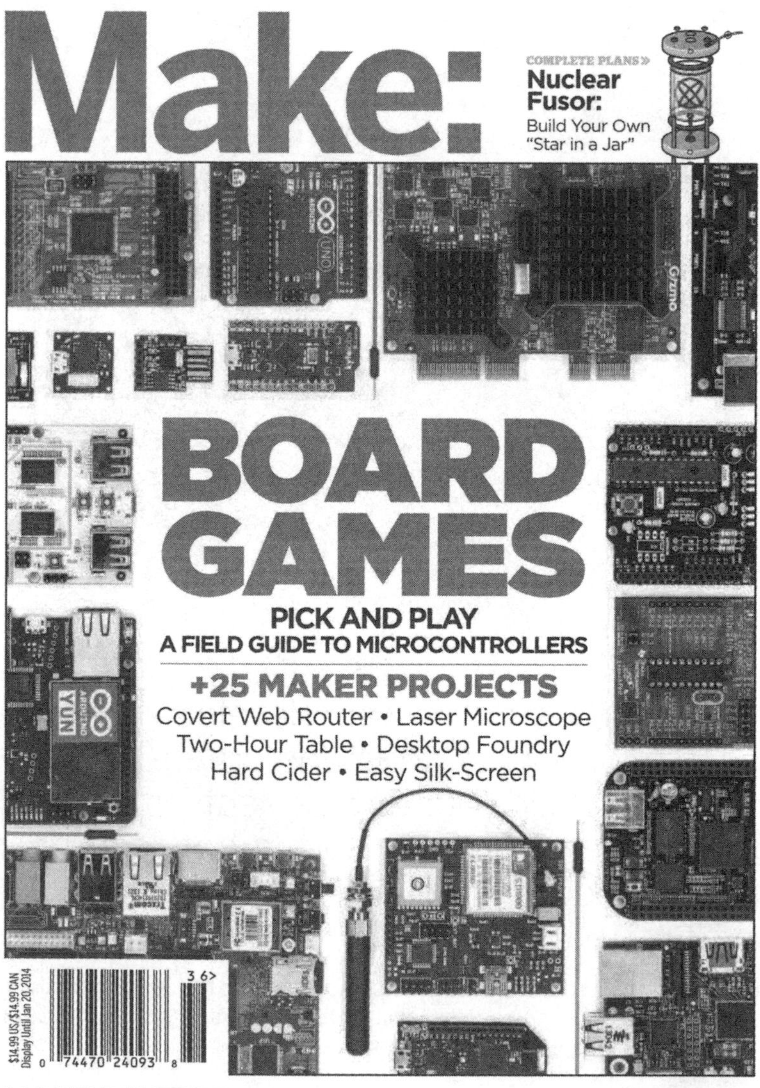

Why Galileo?

When Intel announced Galileo at Maker Faire Rome in October of 2013, there was already an abundant selection of hardware development boards to choose from. At the time, there were so many boards available that an issue of MAKE magazine (Volume 36, Board Games) was released to take on the task of featuring the most interesting boards and helping readers choose the right one for them.

"We're now seeing an explosion of new boards coming to market," wrote Alasdair Allan in that issue of MAKE. "And there's no reason to expect the trend to slow in the next year or two." With so many boards out there, why did Intel decide to jump into this market?

After the announcement of Galileo, Intel CEO Brian Krzanich explained why Galileo came to be. "We wanted to be part of the Arduino ecosystem and maker community for two reasons," said Krzanich to Maker Media's founder, Dale Dougherty. "One was the pure innovation we see happening in the maker community around open source hardware, and we needed to be part of that innovation. Second, we saw that, in education, engineers and others were learning on non-Intel platforms and we wanted to change that, and in doing so, give them more capabilities."

Like the Galileo, the development boards that were gaining popularity had fairly powerful processors, similar to those found in cell phones and tablet computers. What they typically didn't have was an easy-to-use development environment, a good out-of-box experience, or an established community of users. With its strong Arduino compatibility, Galileo excels in these realms. Galileo also gives you the power of Linux under its hood.

Linux is a free and open source operating system that many people run on their desktops and on servers. It's also used in many consumer electronic devices. There's a lot to understand about Linux, but with Galileo, you can focus on bringing your creation to life without needing to know that Linux is there. This makes it easy for users to get more power and capabilities without sacrificing ease-of-use or community support. As you'll see later in this book, you can do some amazing things by poking around under the hood.

Intended Audience

The purpose of this book is to get you started with creating your own hardware projects with Intel Galileo. You won't need any experience wiring up circuits or writing code, but basic computer skills will be helpful so that you can move files around and install the software you'll need to develop projects.

Getting Started with Intel Galileo is written to give you a wide variety of experience and a basic understanding of the many different capabilities of Galileo. It won't dig into electrical engineering or computer science theory. I'll leave that for you to learn elsewhere should you want to pursue those subjects in depth. Instead, I'll focus on how to get things done so that you can experiment, be creative, and make cool stuff with Intel Galileo.

Feedback

I encourage you to contact me with any feedback as you read this book. I hope to be able to incorporate your suggestions into future editions. My email address is *mattr@makezine.com*. You can also find me on Twitter with the name @MattRichardson (*https://twitter.com/MattRichardson*).

Conventions Used in This Book

The following typographical conventions are used in this book:

Italic
: Indicates new terms, URLs, email addresses, filenames, and file extensions.

`Constant width`
: Used for program listings, as well as within paragraphs to refer to program elements such as variable or function names, databases, data types, environment variables, statements, and keywords.

`Constant width bold`
: Shows commands or other text that should be typed literally by the user.

`Constant width italic`
: Shows text that should be replaced with user-supplied values or by values determined by context.

This element signifies a tip, suggestion, or a general note.

This element indicates a warning or caution.

Using Code Examples

Supplemental material (code examples, exercises, etc.) is available for download at *https://github.com/mrichardson23/GSW-Intel-Galileo*.

This book is here to help you get your job done. In general, you may use the code in this book in your programs and documentation. You do not need to contact us for permission unless you're reproducing a significant portion of the code. For example, writing a program that uses several chunks of code from this book does not require permission. Selling or distributing a CD-ROM of examples from MAKE books does require permission. Answering a question by citing this book and quoting example code does not require permission. Incorporating a significant amount of example code from this book into your product's documentation does require permission.

We appreciate, but do not require, attribution. An attribution usually includes the title, author, publisher, and ISBN. For example: "*Getting Started With Galileo* by Matt Richardson (Maker Media) Copyright 2014, 978-1-4493-4537-2."

If you feel your use of code examples falls outside fair use or the permission given here, feel free to contact us at *bookpermissions@makermedia.com*.

Safari® Books Online

Safari Books Online is an on-demand digital library that lets you easily search over 7,500 technology and creative reference books and videos to find the answers you need quickly.

With a subscription, you can read any page and watch any video from our library online. Read books on your cell phone and mobile devices. Access new titles before they are available for print, get exclusive access to manuscripts in development, and post feedback for the authors. Copy and paste code samples, organize your favorites, download chapters, bookmark key sections, create notes, print out pages, and benefit from tons of other time-saving features.

Maker Media has uploaded this book to the Safari Books Online service. To have full digital access to this book and others on similar topics from MAKE and other publishers, sign up for free at http://my.safaribooksonline.com (*http://my.safaribooksonline.com/?portal=oreilly*).

How to Contact Us

Please address comments and questions concerning this book to the publisher:

MAKE
1005 Gravenstein Highway North
Sebastopol, CA 95472
800-998-9938 (in the United States or Canada)
707-829-0515 (international or local)
707-829-0104 (fax)

MAKE unites, inspires, informs, and entertains a growing community of resourceful people who undertake amazing projects in their backyards, basements, and garages. MAKE celebrates your right to tweak, hack, and bend any technology to your will. The MAKE audience continues to be a growing culture and community that believes in bettering ourselves, our environment, our educational system—our entire world. This is much more than an audience, it's a worldwide movement that Make is leading—we call it the Maker Movement.

For more information about MAKE, visit us online:

MAKE magazine: *http://makezine.com/magazine/*
Maker Faire: *http://makerfaire.com*
Makezine.com: *http://makezine.com*
Maker Shed: *http://makershed.com/*

We have a web page for this book, where we list errata, examples, and any additional information. You can access this page at:

http://oreil.ly/getting_started_with_galileo

To comment or ask technical questions about this book, send email to:

bookquestions@oreilly.com

Acknowledgements

I'd like to thank a few people who have provided their knowledge, support, advice, and feedback to *Getting Started with Galileo*:

Larry Barras
Julien Carreno
Michael Castor
Jez Caudle
Pete Dice
Seth Hunter
Tom Igoe
Brian Jepson
Jerry Knaus
Eiichi Kowashi
Mike Kuniavsky
Michael McCool
Jay Melican
Eric Rosenthal
Andrew Rossi
Mark Rustad
David Scheltema
Jim St. Leger

1/Introduction to Galileo

The purpose of the hardware and software that make up the Arduino platform is to reduce complexity when making an electronic project. It's meant to let you experiment, invent, and explore creative uses of technology rather than getting bogged down in technical mastery. By offering compatibility with Arduino hardware and software, Intel Galileo delivers an easy-to-use platform but has more power and features than typical Arduino boards.

What Is Galileo?

Galileo is a *hardware development board*, which is an electronic circuit board that helps you develop interactive objects by reading information from the physical world, processing it, and then taking action in the physical world. If it's connected to a network, it can also communicate to other devices like web servers. Galileo is an Arduino-compatible development board.

What Is Arduino?

There are a few answers to the question, "What is Arduino?" First and foremost, it's a hardware development board like Intel's Galileo. There are models of boards such as the Arduino Uno, Arduino Mega, and Arduino Yún. Each of these Arduino boards has different capabilities. The most basic board, the Arduino Uno, is typically what people are referring to when they say "an Arduino."

There's also the Arduino IDE software, which is the computer application that you use write code and upload it to the board. Arduino is also the name of the language used to program the board.

If you're entirely unfamiliar with Arduino and want to learn more about it, the Arduino website (*http://arduino.cc/*) has many resources including getting started guides, reference information, communities, projects, and news updates. The book *Getting Started with Arduino* by Massimo Banzi (O'Reilly) was my first guide to using the popular development board. It covers the design philosophy of Arduino ("The Arduino Way") and walks you through the basics of using it. This book will cover a lot of that ground as well, but tailored for the Galileo board.

Galileo is an *Arduino-compatible* board, meaning that it can be programmed with the Arduino IDE using the Arduino programming language. It's also compatible with the Arduino 1.0 pinout, the design specification that says which pins go where on the board. Because it's compatible with the Arduino 1.0 pinout, you're able to attach most *Arduino shields*. A shield sits on top of the board and expands the functionality of it. Common circuits to drive motors, control many LEDs, or play sounds can come in the form of shields. The pin layout compatibility also makes it easy to use Galileo when you're following tutorials written for the other Arduino boards.

Inputs and Outputs

Like other hardware development boards, Galileo reads inputs and can control outputs. An *input* brings information from the physical world into the board's processor. It can be as simple as the state of a button or switch but can also be the position of a dial or slider like you see on a sound mixing board. Sensors can also be used as inputs (see Figure 1-1) to read information from the physical world. There are plenty of sensors to choose from including temperature, light level, sound level, acceleration, and much more.

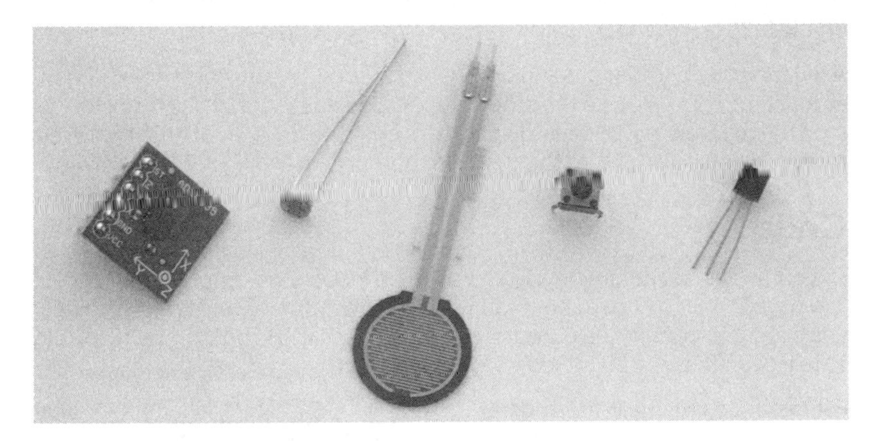

Figure 1-1. *A few possible inputs. From left to right: an accelerometer, a photo cell, a pressure sensor, a button, and a temperature sensor.*

An *output* is how a development board like the Galileo can affect the physical world. It can be as simple as a *light emitting diode*, or LED, which glows when electrical current runs through it. An LED might indicate whether the device is turned on, or if there's an error (a blinking red LED would be perfect for that). Outputs could also be motors that drive wheels on a robot, a text display for the temperature, or a speaker that plays musical tones. Figure 1-2 shows a few.

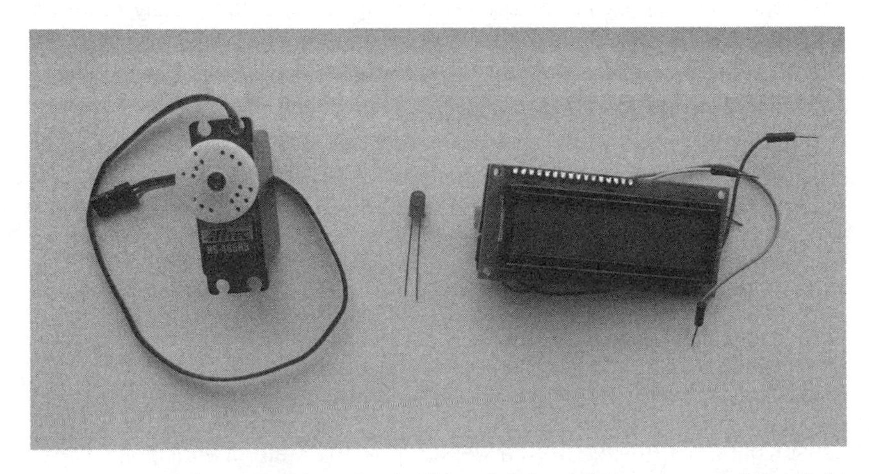

Figure 1-2. *A few possible outputs. From left to right: a servo motor, a light emitting diode, and an LCD character display.*

For example, a simple stopwatch has inputs and outputs. The start button would be considered an input. When you press the start button, it triggers a timer that keeps track of the time and outputs that information to the display on the face of the watch.

A digital voice recorder has a microphone for sound input, and a small speaker for sound output. Like the stopwatch, it also has input buttons to start or stop the recorder and a small display to output the amount of time that's left to record before you fill up the device's memory.

Code

Of course, it's not as simple as just wiring up inputs and outputs to a Galileo. You have to tell the board how you want it to respond to the inputs and how you want it to control the outputs. By programming the board, you'll be able to tell it what you want it to do.

For instance, a simple thermostat project will periodically check the value from a temperature sensor and compare it to the desired temperature that the user set using a dial control. If the temperature that the sensor measures is lower than the desired temperature, the board will activate a heater until the temperature gets close enough to the desired temperature. Logic like this will be defined by the code you write.

The Galileo can be programmed and reprogrammed over and over again. In fact, in the course of developing a project, you'll likely go through a cycle of writing code, uploading it to the board, checking how it works, finding problems, making adjustments to your code, and then uploading it again.

You may even find yourself using the board for one project, and then pulling the board out, reprogramming it, and using it for a completely different project a few weeks later.

Communication

The Galileo can also communicate with other devices in a few different ways. You can have it connect to your computer via USB to send and receive data. You might have Galileo send information about what it's doing to a console window running on your computer so that you can figure out why something isn't working right (this is known as *debugging*). Or you can have it send information about sensors to the computer so that it can display a live graph.

Galileo can also connect to other devices over the Internet using its built-in Ethernet (Figure 1-3) or an optional WiFi module. It can receive information about the weather or your email. It can search Twitter and much more. It can also use the Internet connection to send information such as temperature sensor data, the images from a webcam, or the state of your dog's water bowl.

Figure 1-3. *The Galileo's Ethernet port is just one way it can communicate with users or other devices.*

What Makes Galileo Different?

If you've used a typical Arduino like the Uno before, there are a few key differences between it and the Galileo (Figure 1-4). In fact, the specs on the Galileo make it seem like it's the product of cross-pollination between an Arduino Uno and a low-end computer.

Figure 1-4. *An Intel Galileo next to an Arduino Uno*

The board itself is a little bit larger than an Arduino Uno, but along with that size, you get a more powerful processor (see Figure 1-5), more memory to store running programs, more data storage space, an Ethernet connector for connecting it to a network, and the ability to connect computer accessories through the USB port or the Mini PCI Express connector on the bottom.

Figure 1-5. *Intel's Quark SoC X1000 is the processor at the heart of the Intel Galileo.*

The *firmware* that runs on the Galileo is much more advanced than what's on an Arduino Uno. On an Uno and most other Arduino devices, there is firmware on the board called the *bootloader* which is meant to help you upload and run your code on the board's processor. It only does that and not much else. The firmware on the Galileo, on the other hand, is much more advanced. Not only does it help you upload and run your code on the board, but it also keeps track of files, the date and time of day, and helps share the board's various resources between multiple programs running at the same time. In that way, it's more like a typical computer.

In fact, the firmware on the Galileo is a version of *Linux*, the free operating system that powers many desktop computers and servers these days. Galileo may not have a screen or desktop environment, but it still has much of the functionality that an operating system affords. And through your Arduino code, you'll be able to access this functionality, giving you much more capabilities than you'd have with a typical Arduino. For instance, if you want your project to take a picture from a web cam and email it, it's something that would be difficult to do with only Arduino code. But with the power of Linux, this could be done more easily.

Sketching in Hardware

Artists, engineers, designers, architects, and makers frequently start their work with a simple sketch on paper. Putting the idea down as a sketch helps by pushing something from being abstract towards something more concrete, more real. Sketching something out also helps you communicate your idea to your peers and collaborators. But you don't necessarily need to use a paper and pencil to create a sketch.

Having the power of a computer but the simplicity of Arduino development tools means that there's less to stand between you and your idea for an interactive object. It can help make the abstract idea a little more concrete. As a tool, the Galileo is meant to help you prototype early and iterate often so that you can get an idea of the look and feel of your project and refine it without hassle. I like to call this "sketching in hardware," a term I first heard from Mike Kuniavsky, who organizes a yearly conference by that name. According to Mike, the notion of this term originated with Bill Buxton's work on sketching in the realm of user experience design.

I don't want to be the one to stand between you and sketching in hardware, so let's jump right in.

2/First Steps

Blinking an LED is commonly the first thing to try with a new hardware development board. It's easy to do and it confirms that you have everything working correctly. If you've tried programming before, your first step may have been to get your code to print the text "Hello world." Getting an LED to blink on a hardware development board is its way of saying "Hello world."

By the end of this chapter, you'll learn a few of the different parts of the Galileo, what tools and components you'll need to work with it, and how to install the development software and upload code to the board. To test things out, you'll use Galileo to make an LED blink.

Tour of the Board

First let's take a look at some of the important components on the Galileo. It's not necessary to fully understand what every single part does or how it works in order to get the most out of the board, so I'll just stick to the highlights, which are shown in Figure 2-1.

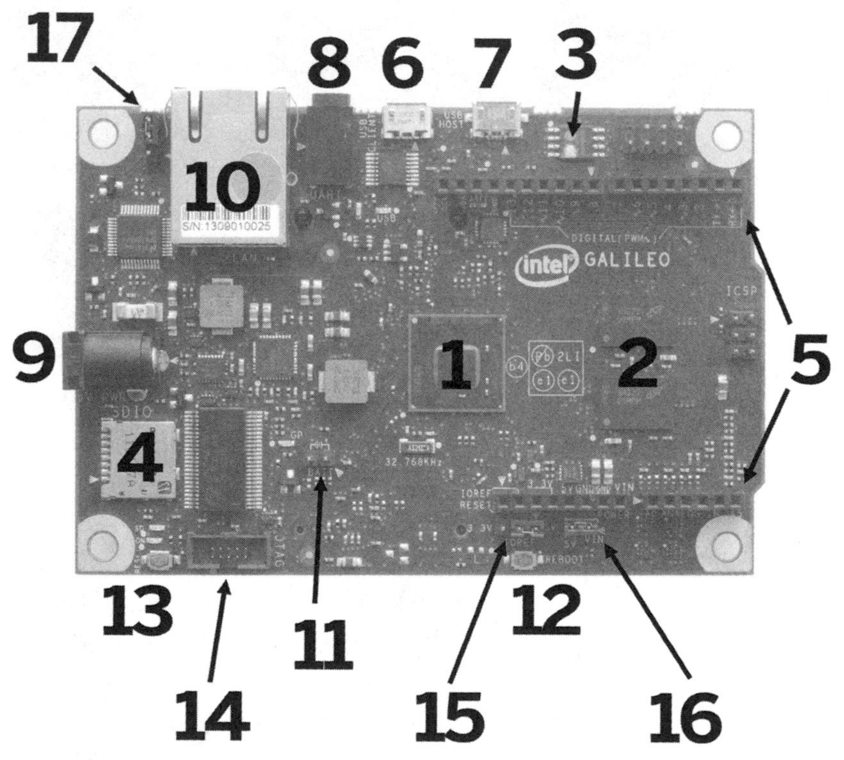

Figure 2-1. *A few of the important components on the Galileo*

Processor (1)

The processor is the brains of the whole operation. Just like the *central processing unit*, or CPU, on your computer, it carries out all the instructions in your program by making calculations and reading or writing data in memory. This particular processor is Intel's Quark SoC X1000 Application Processor, which is designed for small-sized, low-power applications. It's not as powerful as your laptop's CPU, but it's much more powerful than the chip on an Arduino Uno.

Random-access memory (RAM) (2)

Random-access memory, or RAM, is where Galileo keeps running programs and keeps track of data that's being used by those programs. The

Galileo has 512 kilobytes of RAM built into the processor and an additional 256 megabytes of RAM on these chips. When the board is powered down, the data stored in RAM is lost. Any data that should be saved (such as your code) must be stored on the microSD card.

Flash memory (3)
The *flash memory* acts like the hard drive of Galileo. Unlike with RAM, any data stored here is saved even after the board is shut down and the power is disconnected. For this reason, it's where the board's software and operating system are stored. It can hold 8 megabytes of data, most of which is taken up by Galileo's operating system.

MicroSD card slot (4)
If you need more space for larger programs or to store data, you can insert a *microSD card* into this slot. MicroSD cards are just like the memory cards that you insert into a digital camera to save photographs. You can even load an operating system onto the card and boot off of it instead of the on-board Flash memory. You'll need to do this if you want additional functionality like WiFi and access to the webcam because those drivers can't fit onto the 8 megabytes of on-board Flash memory. Galileo can use a card that's up to 32 gigabytes in size. See Appendix D for more information.

Arduino expansion pins (5)
Using these pins, you'll be able to connect to the inputs and outputs on the Galileo. You'll either use *jumper wires* to connect the pins to a *breadboard* for prototyping, or you'll use an Arduino *shield* to add functionality to your board. We'll cover these components a bit more in "Helpful Tools and Components" on page 13.

USB client port (6)
You'll use this port to connect your Galileo to the USB port on your computer. Once it's connected, you can upload your code and communicate with it. Always connect the power supply before plugging the Galileo into your computer over USB.

USB host port (7)
This port allows you to connect USB computer peripherals to your Galileo. It could be accessories such as webcams, sound devices, storage, and much more.

Serial port (8)
This may look like a headphone jack, but it's not meant for audio. It's actually a *serial port*, used for interacting with the Galileo's Linux operating system via a text-based command line environment. See Appendix H for more information.

Power input (9)
This is where you'll plug in Galileo's power adapter. You must plug in the AC adapter when using Galileo. You must always power the board

through its power supply before connecting it via USB to your computer. Otherwise, you may damage your board.

Ethernet port (10)
The Ethernet port on the board will let you connect it to a wired network so that it can communicate with other computers and devices on the network, or access the Internet.

Mini PCI express slot (not pictured)
If you want to make your network connection wireless, you can connect a WiFi card to the *Mini PCI Express slot* on the bottom of the board. This slot can also accommodate cards that offer additional functionality such as more storage space, GSM access for connecting to cellular networks, Bluetooth for wireless device connectivity, and much more.

Clock battery power (11)
This connector will let you wire up a small 3-volt coin cell battery to the Galileo so that the processor can keep track of the date and time even when the board is not connected to 5 volts.

Reboot button (12)
This button will reboot the board, including Galileo's Linux operating system.

Reset button (13)
This button will restart your code and send the reset signal to any shield attached to the expansion header. Galileo's Linux operating system will remain running as normal and won't restart.

JTAG header (14)
This 10-pin connector is mostly used by electrical engineers or advanced hobbyists to test and debug boards.

IOREF jumper (x3) (15)
This jumper lets you change the logic voltage level of the board from 5 volts to 3.3 volts for compatibility with 3.3-volt shields and components. Throughout this book, you'll be using Galileo in its 5-volt mode.

VIN jumper (16)
Pulling this jumper out will disconnect the VIN pin from Galileo's 5-volt regulator. If you're using a shield that requires more than 5 volts on the VIN pin, you must pull out this jumper to protect the board from damage.

I2C jumper (17)
This jumper lets you change the I2C address of a couple of on-board components. You may need to do this if you're using I2C components that conflict with the components on the board. In all likelihood, you won't need to use this.

Helpful Tools and Components

There are a few accessories that you'll want to have handy in order to get the most out of experimenting with your board. If you bought Galileo as part of a bundle, you probably already have a lot of these components (Figure 2-2). You might have others already lying around.

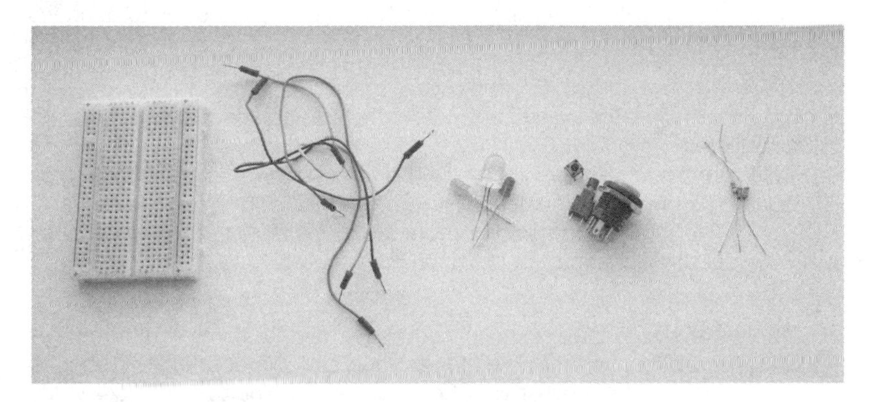

Figure 2-2. *Some very basic prototyping components. From left to right: a breadboard, jumper wires, light emitting diodes (LEDs), buttons, and resistors.*

At the very least, in order to turn on the board and upload code, you'll need:

Computer
: You will need to download the software for programming the board later in this chapter. It can be run on most Windows, OS X, or Linux machines.

Power supply
: The power supply is typically included in the box with your Galileo. If you don't have one, you'll want to get a 5-volt DC power supply that's capable of providing at least 2 amperes. To connect it to your board, it should have a 2.1mm DC barrel connector with a center-positive plug. Most DC barrel connectors are center-positive, but it's good to be sure. (Adafruit.com part number 276)

USB A to micro B cable
: This is the same type of USB cable that is used to connect newer USB devices like Android phones. (Monoprice.com product number 5137, Adafruit.com product number 592, Sparkfun.com product number 10215)

With those basics, you can boot Galileo and upload code to the board. But without a few extra components, you won't be able to make it interact with the physical world. The parts below will be used in the projects and exercises in this book.

Solderless breadboard
: These inexpensive boards are great for prototyping electronics because they make it easy to make connections between components. (Makershed.com part number MKEL3, Adafruit.com part number 64, Sparkfun.com part number 12002)

Jumper wires
: To connect Galileo's expansion pins to the breadboard, or to connect components to each other on the breadboard, you'll use basic male-to-male jumper wires. (Makershed.com part number MKSEEED3, Adafruit.com part number 758, Sparkfun.com part number 08431)

LEDs, assorted
: LEDs are a part that I reach for all the time when I'm experimenting or building a project. I usually buy them in three colors: red, amber, and green. (Makershed.com part number MKEE7, Adafruit.com part number 299, Sparkfun.com part number 12062)

Resistors, assorted
: These are inexpensive electronic components that come in different resistance values. Vendors like Maker Shed and Jameco sell multi-packs of resistors that contain a variety of values. Any multipack of 1/4-watt, 5%-tolerance resistors will be perfect to get started. (Makershed.com part number MKEE4, Sparkfun.com part number 10969)

Buttons and switches, assorted
: These are the kind of components that you can harvest from old electronics and appliances. Or you can stop by your local RadioShack and browse the selection in their component bins, which is one of my favorite things to do! (There are a huge variety of buttons and switches. For the most basic type of switch that will fit in a breadboard: Adafruit.com part number 00097, Sparkfun.com part number 00097)

MicroSD card
: Galileo has a limited amount of on-board storage. With a MicroSD card, you can boot from a version of the Linux operating system that has more features and can store data from your projects. Galileo can use a card that's up to 32 gigabytes in size. (These days you can get MicroSD cards at supermarkets, drugstores, or any place that sells electronics.)

USB OTG adapter
: If you want to connect USB devices to your board, you'll probably need a USB OTG adapter with a USB A female end (which is where a USB device will plug in). I've found them to be as cheap as $3.50 online (see Figure 2-3). (Monoprice.com part number 9724, Adafruit.com part number 1099, Sparkfun.com part number 11604)

Figure 2-3. *A USB OTG adapter like the one you see here will help you connect USB components to the Galileo.*

These parts are nice to have, but there's no need to rush out and buy them:

Serial cable
: Using a serial cable, you can access the Galileo's Linux text-based command line environment from your computer. There are other ways to get to this command line without the cable, so this will only be needed if you run into trouble. See Appendix H for more information.

Case
: A case isn't required, but it may help to make sure that your board is protected from the hazards on a maker's workbench such as spilled beverages. If you have access to a 3D printer, you can even print your own (*http://www.thingiverse.com/thing:159983*)! Engrained Products (*http://engrainedproducts.com/*) was one of the first companies to sell cases made specifically for Galileo.

PowerSwitch Tail II
: This handy device will help you use the Galileo to control A/C appliances like lamps and blenders. (Makershed.com part number MKPS01, Adafruit.com part number 268, Sparkfun.com part number 10747)

Writing Programs to Control Your Galileo

In order to write the code to program the board, you'll need the Arduino *integrated development environment*, or IDE, on your computer. It's also sometimes referred to as the Arduino software. The development environment is the program where you'll write your code, check it for errors, and upload it to the Galileo. To download the software:

1. Navigate to the Galileo support site (*http://www.intel.com/support/galileo/*) in your web browser.
2. Click "Software downloads."
3. There will be a list of different files to choose from. Find the file name that starts with *Intel_Galileo_Arduino_SW_on* and then your operating system. It'll be either *Linux32bit*, *Linux64Bit*, *MacOSX*, or *Windows* depending on what system you're using.
4. Click the download button for that file.
5. After accepting the terms in the license agreement, your download will start.

Because the Galileo support site and the IDE software frequently change, these directions may differ slightly from your experience.

Next, install the software on your computer:

1. Open the file that you've downloaded to decompress it. Find your downloads folder and double-click the file to do so. Some web browsers will also let you open the file by clicking on the download directly in the browser window.
2. From here, there are some differences in how each operating system installs and launches software. Here's a basic overview of the installation and launching process for each platform:
   a. **On Windows**, unzip the file to the root of your hard drive (`C:\`). Open that folder and double click `arduino.exe` to launch the IDE. You will need to install serial drivers to communicate with the board. See Appendix E for more detailed instructions and notes.
   b. **On Linux**, unzip the file and change (cd) into the newly-created directory. Run the Arduino executable from the command line. See Appendix F for more detailed instructions and notes.

c. **On Mac OS X**, simply drag the Arduino icon into your `/Applications` folder to install it. Launch it by double-clicking the Arduino icon in your `Applications` folder. See Appendix G for more detailed instructions and notes.

For OS X systems, if you already have a copy of the standard Arduino IDE installed, you can rename the one you downloaded from Intel and keep both in your `Applications` folder. The name you choose must not contain spaces, so just `Galileo` would be a good choice.

Getting Familiar with the Development Environment

When you open the IDE for the first time, you'll be presented with a window for a new sketch (Figure 2-4). A *sketch* is an Arduino project's code files. Here are a few of the parts of the IDE.

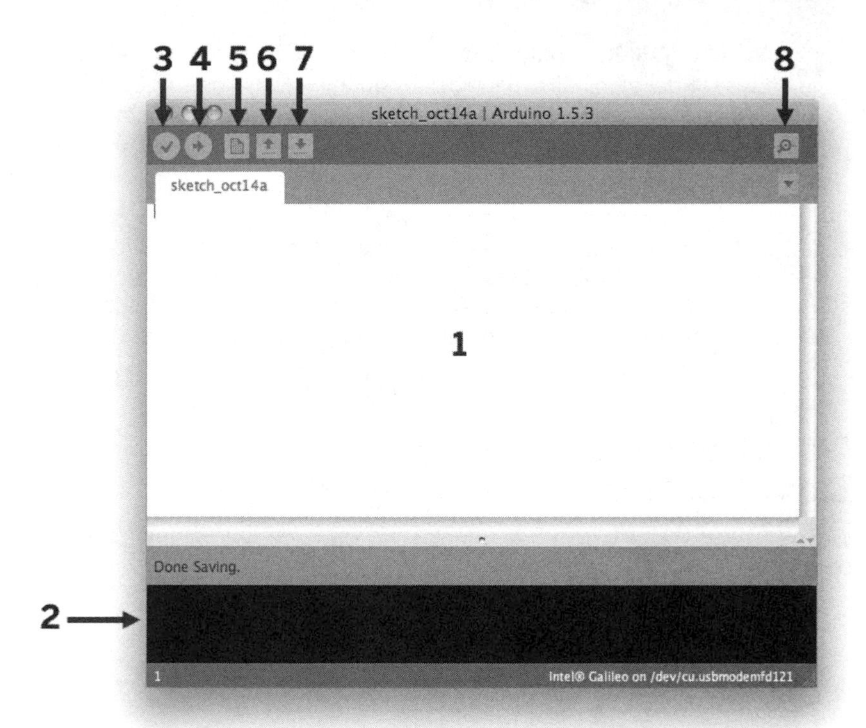

Figure 2-4. *The Arduino IDE is where you'll write code and upload it to Galileo.*

Code entry area (1)
: This is where you'll type in the code for your sketch.

The console (2)
: This is where you'll see status messages when uploading your code to the board.

Verify button (3)
: Click this button to verify that your code has no errors. If your code does have errors, they will be displayed in the console.

Upload button (4)
: Click this button to verify the code and upload it to the board if there are no errors.

New button (5)
: This creates a new sketch in the current window.

Open button (6)
: This will allow you to open a saved sketch from your *sketchbook folder*, the location on your computer where the Arduino IDE saves your sketches.

Save button (7)
: This will allow you to save the sketch in the current window. You'll be prompted to enter a file name if you haven't already saved the file.

Serial monitor (8)
: This button opens the *serial monitor*, which lets you send and receive information between your board and your computer. We'll cover this further in Chapter 3.

You'll get a chance to explore more features of the IDE as you progress through this book.

Connecting the Board

The next step is to power on the Galileo and connect it to your computer (Figure 2-5) so that you can upload your code to the board.

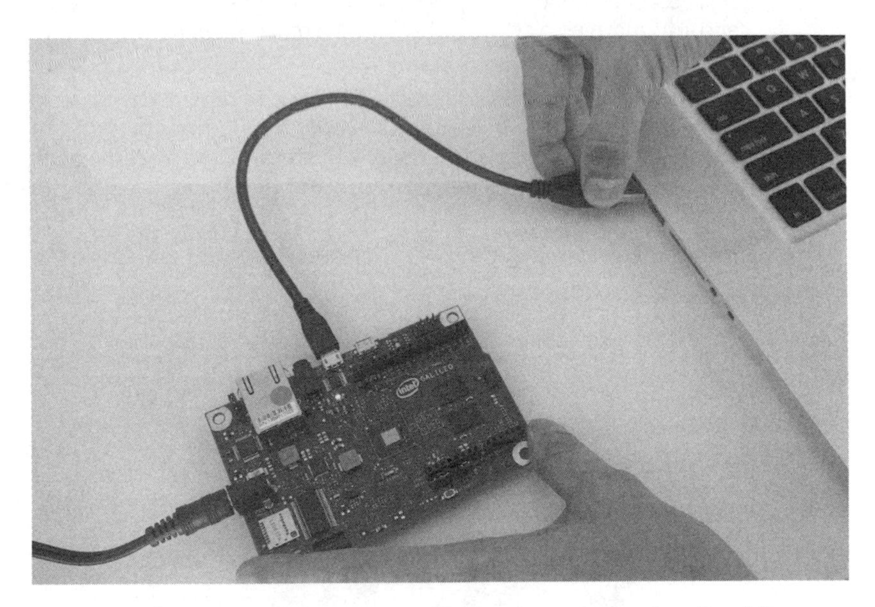

Figure 2-5. *After connecting your Galileo to power using the AC power adapter, plug it into your computer via USB.*

You must always power the board through its power supply before connecting it via USB to your computer. Otherwise, you may damage your board.

1. With nothing else connected to your Galileo, plug the 5-volt power supply into your wall and connect its DC jack to the power jack on the board. You should see a couple of LEDs light up.
2. Then connect Galileo to your computer using a USB A to micro B cable. The cable will connect to the USB client port on the board, which is right next to the Ethernet port. See Figure 2-5.
3. In the Arduino IDE, go to the Tools pull-down menu at the top, select Board, and make sure "Intel Galileo" is checked.
4. Then pull down the Tools menu again, and select Serial Port.
5. Depending on your system, you may see a few different options under the Serial Port menu (Figure 2-6).
   a. **On Mac OS X**, select the serial port that shows something similar to "/dev/cu.usbmodemfd121" (see Figure 2-6). The last few digits will likely be something else. Do not select the one beginning with "/dev/tty".
   b. **On Windows**, select the serial port COMx, with x replaced by a number specific to your system. You can verify the number by going to Control Panel→System and Security→System→Device Manager. In the "Ports (COM & LPT)" section, use the COM port assigned to "Gadget Serial."
   c. **On Linux**, select the serial port that shows something similar to "/dev/ttyAMC0."

Keep in mind that it takes a bit of time for the board to boot up. You may not see the port appear in the menu immediately after powering the Galileo and plugging it into your computer.

If you have multiple serial ports to choose from and aren't sure which to choose, try unplugging the Galileo from your computer, checking the Serial Ports menu, plugging the board back in, and opening the menu again to see which port gets added.

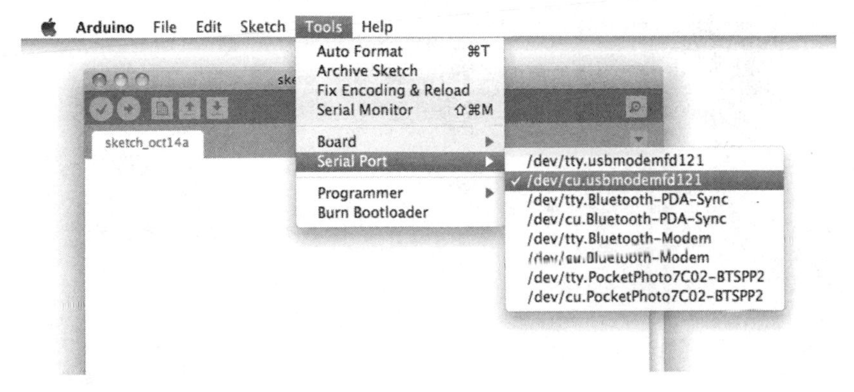

Figure 2-6. *Your serial port will likely have a different name, but on a Mac, it will be similar to the one selected here.*

Now is a good time to make sure that you have the most up-to-date firmware on the Galileo. Click Help→Firmware Update to update your board to the latest version.

Uploading Code

One of the fantastic things about Arduino is the excellent code examples that come along with the IDE. Now you'll open the most basic example and upload it to the board.

1. Within the IDE, click File→Examples→01.Basics→Blink. See Figure 2-7.
2. A new sketch window will open up with some code in it.
3. Click the Upload button on the tool bar.
4. It may take a moment to compile and upload to the board. You'll see text go by in the console at the bottom of the window.
5. When it's done, you'll see the text "Done uploading" at the bottom.

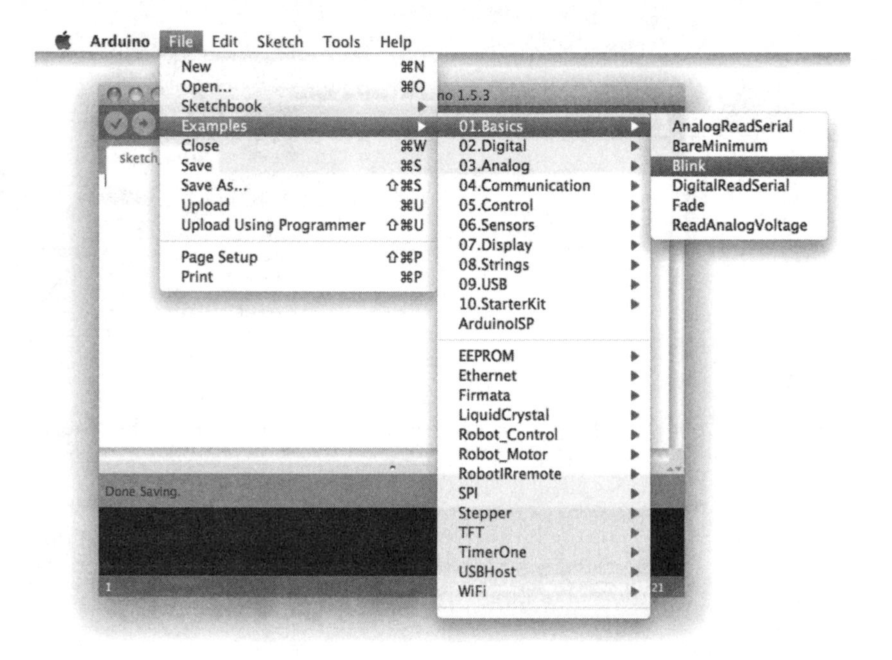

Figure 2-7. *Navigating through the examples menu to open the Blink example*

While you were waiting for "Done uploading" to appear, the Arduino IDE *compiled* the sketch, which means it turned the code into instructions that Galileo's processor will understand. The IDE then uploaded the compiled program to the board.

Now when you look at the board, you should see an LED blinking on and off right next to the clock battery connector (see Figure 2-8). If not, you'll want to troubleshoot this before moving forward. Here are a few troubleshooting steps:

- Make sure the Galileo is connected to power and to your computer via USB.
- Make sure Intel Galileo is selected from the Tools→Board menu.
- Try another serial port in Tools→Serial Port.
- The firmware on your board may not be up to date. Try clicking Help→Firmware Update while the board is connected and powered on.
- Make sure you have the Blink sketch open and you didn't accidentally change any of the code.
- Is there an error message that might give you clues about what's wrong?

If all else fails, you can always seek help from the Galileo Support Community (*https://communities.intel.com/community/makers*) or the Arduino Forums (*http://forum.arduino.cc/*).

Figure 2-8. *The small on-board LED attached to pin 13 will blink after uploading the example code.*

Taking It Further

Throughout the code, there are plain English explanations written as *comments*. Any text between a /* and a */ is considered a comment and is ignored by the Arduino compiler. And on a line of code, any text after a // is also treated as a comment and is ignored. You'll notice that the Arduino IDE sets the color of comments to gray so that it's easy to distinguish from the actual code (see Example 2-1).

Example 2-1. Comments in Arduino code

```
/*
  Everything here is ignored by Arduino's
  compiler.
*/

int led = 13;

void setup() {
  // This is also ignored.
  pinMode(led, OUTPUT);
}

void loop () {
  digitalWrite(led, HIGH); // This text is ignored as well.
  delay(1000);
  digitalWrite(led, LOW);
  delay(1000);
}
```

Take a look at the code in the Blink sketch and see if you can understand what it's doing by reading the comments. We'll dig deeper into the structure of Arduino sketches in Chapter 3.

Try making some changes to the code. For instance, what do you think you should change to make the LED blink faster? Try changing the code and uploading it to the board again.

What happens when you delete a semicolon at the end of a line and try to upload the code?

What happens when you delete one of the curly brackets and try to upload the code?

If you're curious, feel free to explore the other examples that come along with the Arduino IDE!

3/Outputs

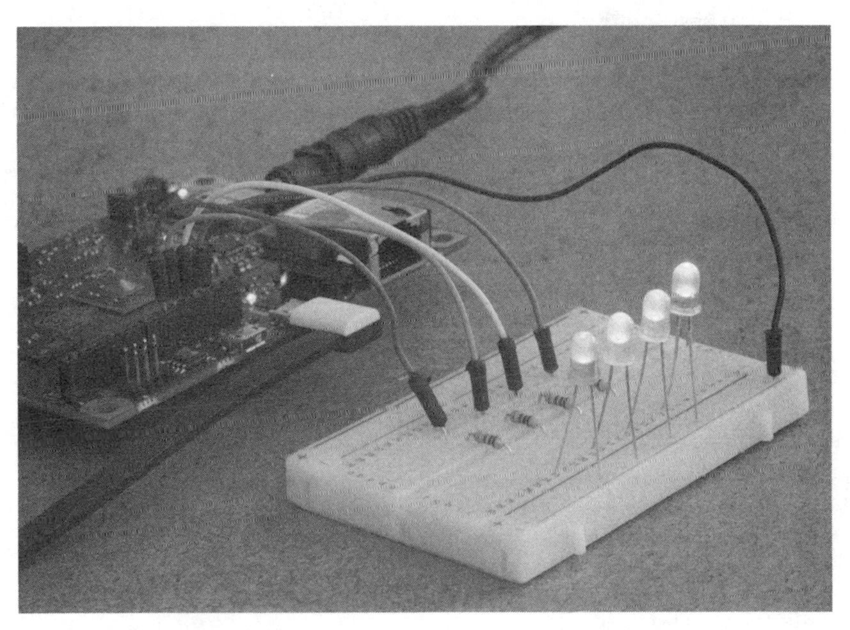

A blinking LED is just one example of an output with Galileo, and there are plenty of other types of outputs available to you. For instance, a speaker can play tones or a motor can drive the wheels of a robot. LEDs, speakers, and motors convert the electrical energy from Galileo into some other form of energy. An LED converts electrical energy into light energy. A speaker converts electrical energy into sound energy. And a motor converts electrical energy into mechanical energy.

Outputs can be used to communicate information to a user, make things move, or send signals to other devices. In this chapter, you will use Galileo's output functionality to:

- Learn basic Arduino syntax
- Experiment with a few different types of outputs
- Learn basic output functions
- Learn about the difference between digital and analog
- Learn how to send information from Galileo to your computer

Back to Blinking: Digital Output

At the end of Chapter 2, you connected Galileo to your computer and uploaded example code that caused an LED to blink on and off. Let's take a closer look at the code, which I've reproduced in Example 3-1.

Example 3-1. The Arduino Blink Example

```
int led = 13;

void setup() {
  pinMode(led, OUTPUT);
}

void loop() {
  digitalWrite(led, HIGH);
  delay(1000);
  digitalWrite(led, LOW);
  delay(1000);
}
```

Setup and Loop

The first thing to notice is that there are two separate blocks of code enclosed within curly brackets. One starts with `void setup()` and the other starts with `void loop()`. Every Arduino sketch you write will have both of these.

The block of code that starts with `void setup()` is the *setup function*. When your sketch starts, the Galileo will execute each line of code within the setup function, starting with the first line and then working its way down. It will then move onto the *loop function*, which is all the code in the curly brackets after `void loop()`. Galileo will repeatedly execute the code in the loop function over and over again until power is shut down or the reboot or reset button is pressed. Again, it will execute the code in the order it is written (see Figure 3-1).

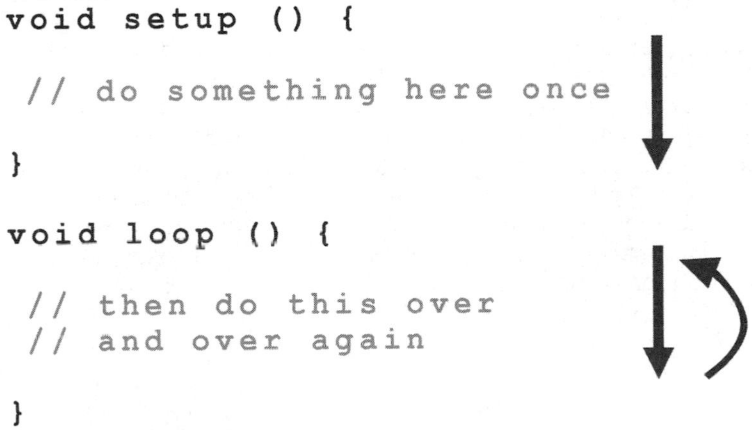

```
void setup () {

  // do something here once

}

void loop () {

  // then do this over
  // and over again

}
```

Figure 3-1. *The structure of every sketch starts with a setup and loop function. The code in the setup function is executed once at the beginning of your sketch, and then code in the loop function executes over and over again*

To summarize, when an Arduino sketch runs, it will:

- Execute the code in the setup function once.
- Then it will execute the code in the loop function over and over again.

Variables

Now you may be eyeing the line that says `int led = 13;` right before the setup function in Example 3-1. That line is creating a *variable*, a place in memory to store data. In this case, you're storing an *integer*, which is a whole number. This line gives this spot in memory a name (`led`) and stores the value 13 in it.

A variable is like a locker to store data. At any point in your sketch, your code can open the locker to access the data inside, or it can replace that data with other data. `int led = 13;` means that your code is creating a locker with the label `led` and it's going to hold integers (as opposed to other types of data such as floating point numbers or text). It's also going to put the value 13 in that locker.

If you jump down further in the code, any spot where you see `led` is where the sketch will be accessing that number 13 from memory.

Pin Numbers

But why store the integer 13 in a variable called `led`? The reason is that there are 14 *digital input/output pins* on the Galileo (see Figure 3-2). These are pins that can be used to control outputs or read inputs. (In this chapter you're using them as outputs only.) These pins are numbered 0 through 13 and you use one of those numbers in your sketch to identify the pin that you want to turn on or off. If you store one of those pin numbers in a variable, you can use that variable name (in this case, `led`) in your sketch instead of having to refer to the pin number (in this case, 13) over and over in your sketch (see "Hard-coding" on page 40).

Figure 3-2. *The fourteen digital input/output pins on the Galileo are numbered 0 through 13.*

Pin number 13 is different from the others because it's wired up to an LED that's attached to the board. That's the one that was blinking after you uploaded the code at the end of Chapter 2.

Pin 13 is also wired up to the tiny hole labeled with the number 13 on the Arduino expansion pins. To prove it, follow the instructions below to wire up another LED.

For this exercise, you'll need:

- Jumper wires (Makershed.com part number MKSEEED3, Adafruit.com part number 758, Sparkfun.com part number 08431)
- Breadboard (Makershed.com part number MKEL3, Adafruit.com part number 64, Sparkfun.com part number 12002)
- An LED (Makershed.com part number MKEE7, Adafruit.com part number 299, Sparkfun.com part number 12062)
- A 330 ohm resistor (Assorted resistor packs: Makershed.com part number MKEE4, Sparkfun.com part number 10969)

Here's how to wire up the LED, step-by-step:

1. Using a jumper wire, connect one of the Galileo pins labeled GND into one of the rows on your breadboard. (If you've never used a breadboard before, review Appendix B.) There are a few pins labeled GND, and any one of them will be fine. This is the LED's connection to *ground*.
2. Using another jumper wire, connect the Galileo pin labeled 13 to a different row on the breadboard as shown in Figure 3-3.

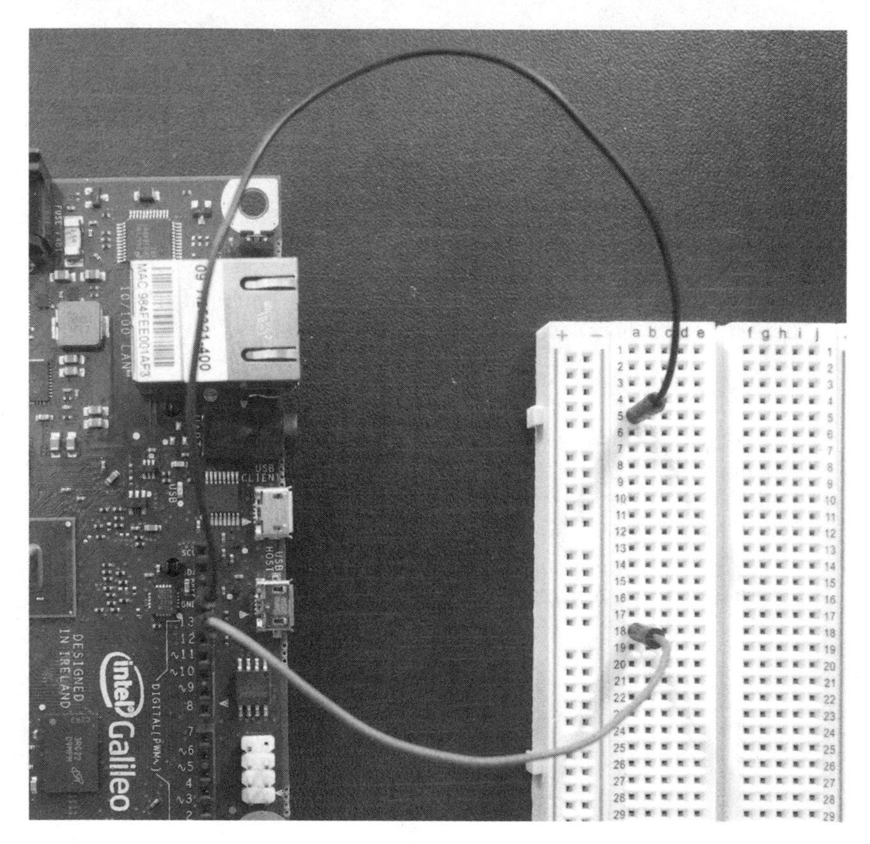

Figure 3-3. *Connecting ground and digital pin 13 to a breadboard with jumper wires*

3. On your LED, identify which wire is the *anode* and which is the *cathode*. The anode (positive) is the longer wire coming off the LED. The shorter wire is the cathode (ground). See Figure 3-4.

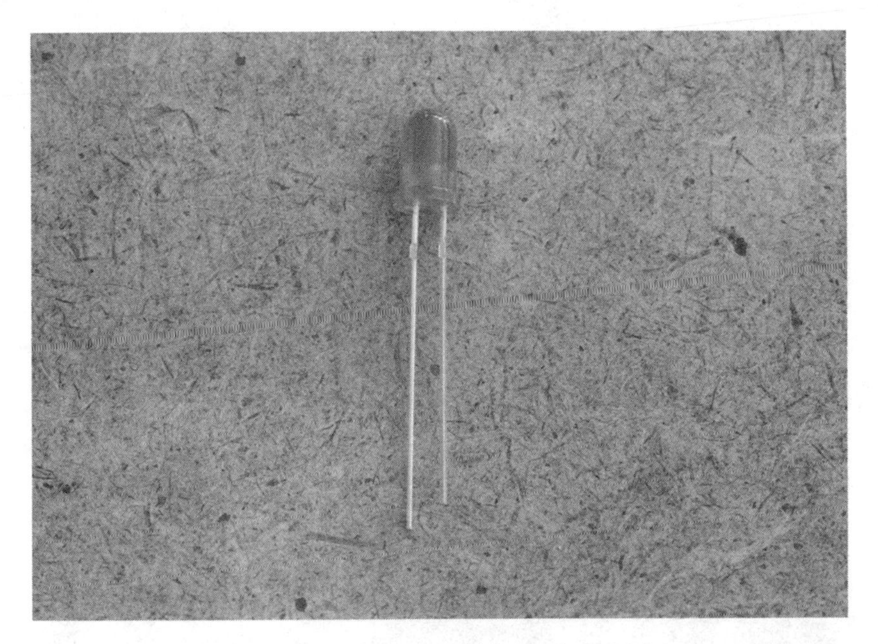

Figure 3-4. *On an LED, the longer lead (on the left side) is the anode. The shorter lead (on the right side) is the cathode. The anode connects to positive voltage and the cathode connects to ground.*

4. Insert the LED's cathode (the shorter wire) into the same row of the breadboard that's connected to ground. The anode (the longer wire) can be inserted into an unused row on the breadboard.
5. To complete the circuit, you need to insert a *resistor* as well, *in series* with the LED. The value of the resistor depends on the characteristics of LED you are using and how much electrical current the Galileo provides through the digital pins (10 milliamperes). For a typical red LED, use a 330-ohm resistor, which has bands colored orange, orange, brown, and then gold or silver. See appendix Appendix C on determining resistor values based on their color bands.

This handy resistor calculator for LEDs (*http://led.linear1.org/1led.wiz*) can help you determine the right resistor to use with your LED. The source voltage will be 5 volts, the diode forward voltage depends on your LED (typically 2 volts for a red LED), and the forward current should be set to 10 mA to match the current that Galileo will provide. Your resistor value doesn't need to be spot on, but get it as close as possible.

6. Insert the resistor between the breadboard row connected to pin 13 and the anode of the LED, completing the circuit. Your circuit should look something like figure Figure 3-5 (take a closer look in Figure 3-6).

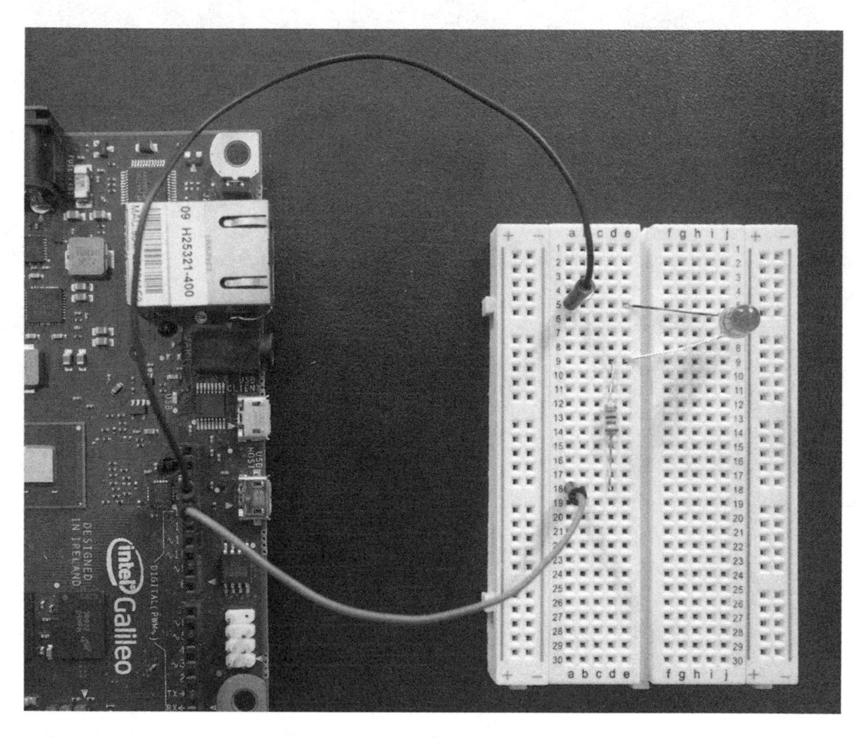

Figure 3-5. *A wire connects Galileo's ground to the cathode side of the LED. The other wire connects Galileo's digital pin 13 to a 330 ohm resistor, which connects it to the anode side of the LED.*

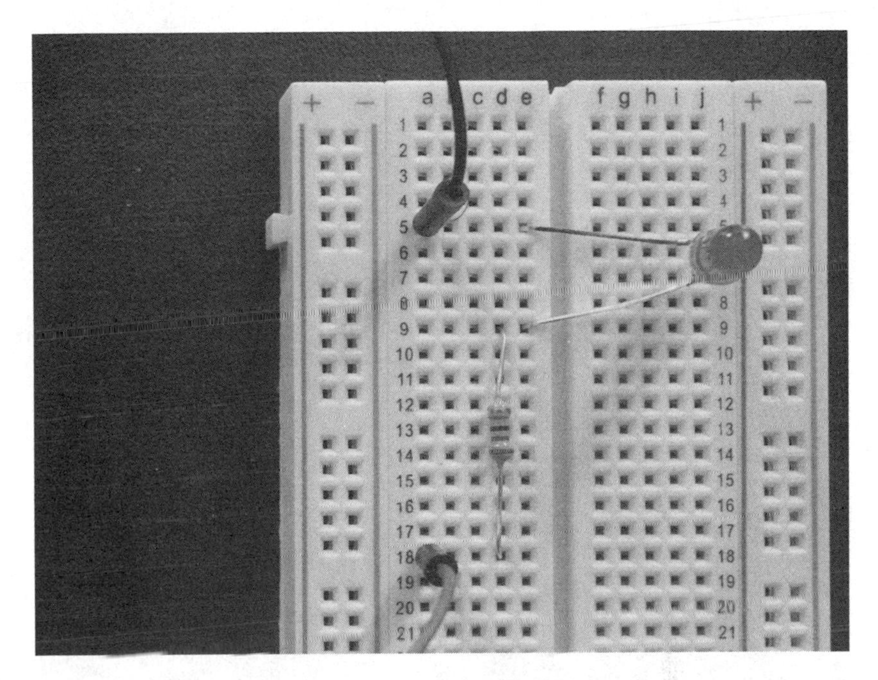

Figure 3-6. *A closer look at the breadboard layout*

As long as you have the Blink sketch ("Uploading Code" on page 21) uploaded, you should see your LED blink along with the on-board LED.

Figure 3-7 shows a simplified diagram of how these parts connect together. Your wires don't necessarily have to be arranged in this orientation.

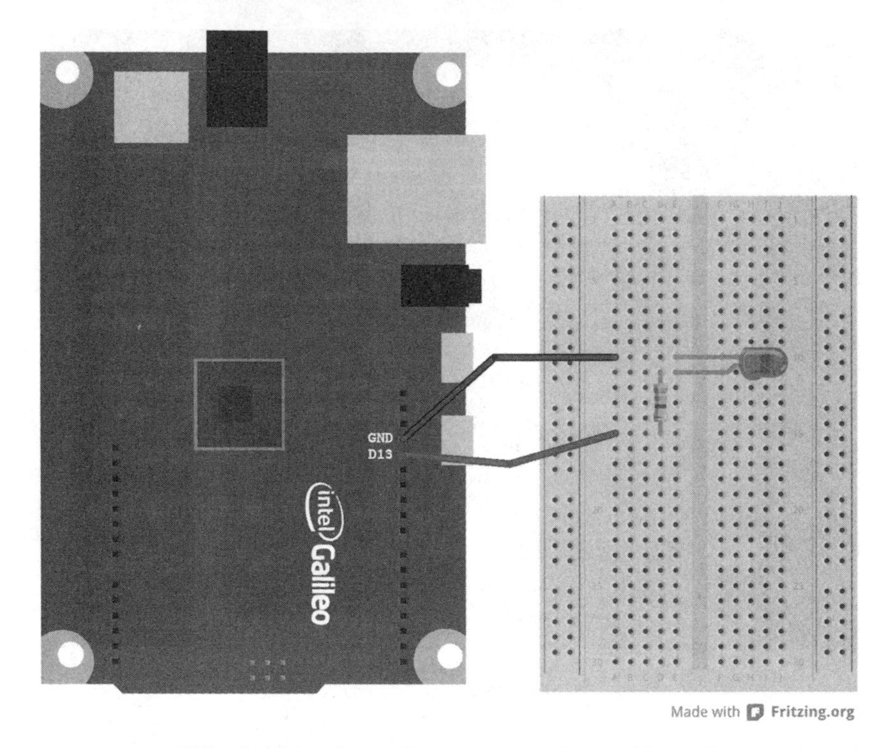

Figure 3-7. *This diagram shows the components used to make an external LED blink: the Galileo, a resistor, and an LED.*

Circuits and the Flow of Electricity

Before jumping back into the code, let's take a break to explain what's going on with the circuit you created. *Electrical current* is the flow of *electrons* through a *conductor*. It's a type of energy. The electrical current from pin 13 (when it's turned on) will move towards the ground pin through a material that allows electrons to move through it. Materials like copper and other metals allow electrons to move freely and are therefore considered good conductors of electricity. That's why many wires are made with copper. The loop that allows the electrical current to flow from the pin to ground is called a *circuit* (see Figure 3-8).

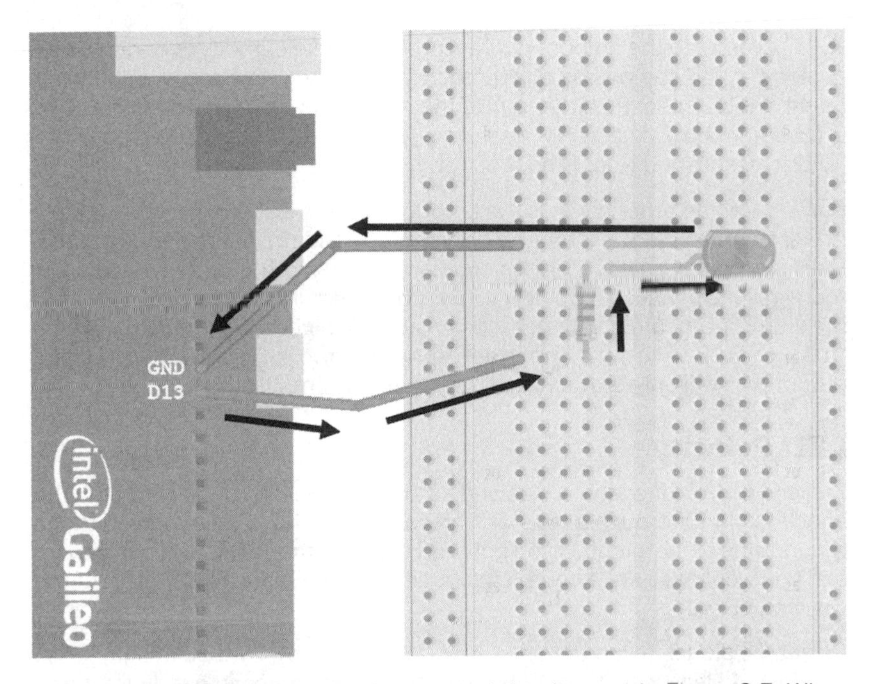

Figure 3-8. *This is a closer look at part of the diagram in Figure 3-7. When making the LED blink, the flow of electric current makes a complete loop (or circuit) from the digital output pin to ground.*

The energy from the flow of electrons in a circuit can be used by the LED, which is converting that electrical current into light energy.

If you were able to see electrons in the wires, the flow that makes up electrical current is actually moving *from* ground *to* pin 13. When we think of the flow of electricity while making circuits, we think in the reverse of reality: that the flow of electrical current goes from a positive charge source (like digital pin 13) to ground. This is by convention only. When electricity was first discovered, early physicists weren't sure which direction electricity flows and so they made a guess. We later found out that they were wrong. However, that early convention is still used.

Most of the time, the circuits you'll be making with Galileo will be using 5 volts. The IOREF jumper, as shown in Figure 2-1, will let you change the logic voltage level for compatibility with shields that use 3.3 volts. Throughout this book, you'll be using Galileo in its 5-volt mode.

Voltage refers to the amount of pressure that the electricity is under. As an analogy, if you had a hose connected to a wimpy little water pump, water might dribble out of the other end of the hose. If you connected a more powerful water pump (that is, increased the voltage), the water would be under more pressure and would shoot out of the hose.

Current, on the other hand, is the amount of electrical energy (the number of electrons) that flows through a given point in the circuit in a period of time. Using the water analogy, attaching a larger hose to the pump would allow more gallons per second to flow past a particular point. Electrical current is measured in *amperes* (amps). A typical red LED might consume 20 milliamps (mA), or 0.02 amps. It's important to note, however, that Galileo's digital output pins will only provide about 10mA of current.

LED Polarity

Earlier, you identified the anode and cathode of the LED and connected the anode to pin 13 through a resistor. The cathode of the LED connected to the ground side of the circuit. If you got this reversed, the LED would not light up and you wouldn't have a complete circuit. This is simply a side effect of the way an LED works.

In fact, all diodes (even the ones that don't light up), by definition, oppose the flow of electricity in only one direction. This comes in handy for some circuits where you need to protect sensitive components from a possible condition in which the flow of electricity reverses.

A component that must be connected in a certain orientation in relation to the flow of electrical current is considered to be *polarized* or have *polarity*.

Resistors with LEDs

Another property of an LED is that it has a typical voltage and current at which it's meant to operate. Differences in color, size, and manufacturer can have an effect on these properties. If the LED is given too much voltage or is allowed to pull more current than it needs, it will be damaged or burn out.

The purpose of a resistor is to resist the flow of electrical current. Using the water hose analogy, it would be like stepping your foot down on the hose. By adding a little bit of resistance to your LED circuit, you can ensure that the LED is provided with the proper voltage and current.

Unlike LEDs, resistors do not have polarity, so the orientation of this component in relation to the flow of electrical current doesn't matter.

That's enough discussion of electrical matters for now. Let's get back into the code.

pinMode()

The only line of code in the Blink sketch's setup function says `pinMode(led, OUTPUT);`. `pinMode()` is one of many Arduino *functions* available to you. A function is a collection of Arduino statements. You run the statements inside a function when you *call* it by name. Functions each have a particular task that they do. They sometimes take data inputs, which are called *parameters*. They sometimes can give back data called *return values*. To understand these concepts better, let's take a closer look at the use of `pinMode()` in the Blink sketch.

By *calling* `pinMode()`, you're telling Galileo that you're going to use a particular pin either as an input or an output. Therefore, it'll need two pieces of information: the pin number and the mode (input or output). You need to do this for each digital pin you plan to use before you use it. Typically, you'll do this in the setup function.

Every function's parameters are passed to it by entering the values in parentheses after the name of the function. If there are multiple parameters, each value is separated by a comma. In the case of `pinMode()`, the first parameter is the number of the pin you want to set the mode of and the second value is the pin's mode, written as INPUT or OUTPUT (in all caps).

To summarize, the *syntax* of `pinMode()` is:

```
pinMode(pin, mode);
```

The parameters of `pinMode()` are:

- `pin`: the pin number
- `mode`: either INPUT or OUTPUT

`pinMode()` doesn't return a value. In Chapter 4, you'll take a look at functions that return values.

In the Blink sketch, the statement `pinMode(led, OUTPUT);` takes the pin number value from the variable `led` (in this case, 13) and sets that pin as an output.

Arduino Language Reference

But how would you know this information if you didn't already have a book or teacher walking you through this? The Arduino Language Reference (*http://arduino.cc/en/Reference/HomePage*) lists each Arduino function, its purpose, parameters, and return values. Each even has an example to demonstrate how to use it (see Figure 3-9).

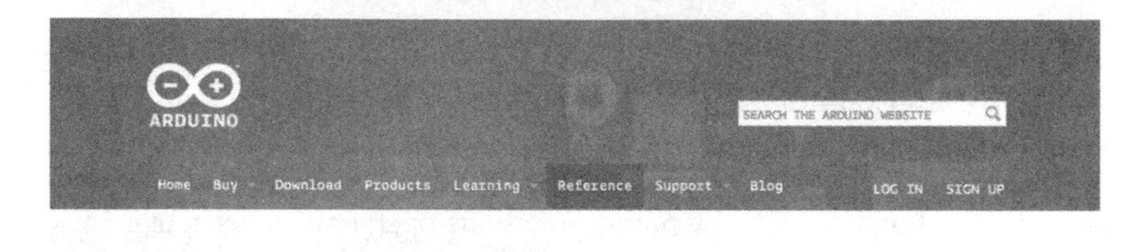

Reference Language | Libraries | Comparison | Changes

Language Reference

Arduino programs can be divided in three main parts: *structure*, *values* (variables and constants), and *functions*.

Structure

- setup()
- loop()

Control Structures

- if
- if...else
- for
- switch case
- while
- do... while
- break
- continue
- return

Variables

Constants

- HIGH | LOW
- INPUT | OUTPUT| INPUT_PULLUP
- true | false
- integer constants
- floating point constants

Data Types

- void
- boolean
- char
- unsigned char

Functions

Digital I/O

- pinMode()
- digitalWrite()
- digitalRead()

Analog I/O

- analogReference()
- analogRead()
- analogWrite() - *PWM*

Due only

- analogReadResolution()

Figure 3-9. *The Arduino Language Reference page (http://arduino.cc/en/Reference/HomePage) is where you can find information about all of the available Arduino functions.*

If you look at the reference page for pinMode() (*http://arduino.cc/en/Reference/PinMode*), the information there should now look familiar to you. It tells you what `pinMode()` does, shows its syntax, explains the parameters, and notes that it does not return a value.

digitalWrite()

Within the loop function, the Arduino function `digitalWrite()` is called twice. `digitalWrite()` turns a pin on or off (*high* or *low* in digital electronics parlance). When the pin is set to high, 5 volts flows through the pin and can be used to power the LED. When the pin is low, it's connected to ground and the LED will turn off.

The syntax of `digitalWrite()` is:

```
digitalWrite(pin, value);
```

The parameters of `digitalWrite()` are:

- `pin`: the pin number
- `value`: either HIGH or LOW

`digitalWrite()` does not return a value. Again, I'll cover return values in Chapter 4.

Therefore, within the Blink sketch, `digitalWrite(led, HIGH);` sends 5 volts to pin 13, illuminating the LED.

And `digitalWrite(led, LOW);` connects pin 13 to ground, turning off the LED.

delay()

Your code will run *very* quickly, so if you only turned the pins on and off in the loop function, there won't be enough time to see the LED fully on or fully off and it'll simply appear dim. Therefore, you need to tell Galileo to wait for a second after turning it on and again after turning it off.

To do that, you'll use the Arduino function `delay()`, which stops your program in its tracks for the amount of milliseconds that you specify.

The syntax of `delay()` is:

```
delay(ms);
```

The parameter of `delay()` is:

- `ms`: the number of milliseconds to wait

`delay()` does not return a value.

In the case of the Blink sketch, `delay(1000);` tells Galileo to wait for 1 second (1,000 milliseconds) after setting pin 13 high and again after setting it low.

Code and Syntax Notes

Now that you've walked through each line of code in the Blink sketch, let's discuss some of the details when it comes to code.

Semicolons

You probably noticed that many lines of code have a semicolon at the end. This is called a *terminator* and it lets the Arduino compiler know that it has reached the end of a *statement*. In programming, a statement is like a sentence and in Arduino syntax, the semicolon is like the period at the end of a sentence.

There are some cases when you don't use a semicolon, such as when you're opening a block of code. You'll see in this in the Blink sketch with the lines that start the setup and loop functions. You also do not need a semicolon after the curly bracket that closes a block of code.

Lines and Spacing

Because the Arduino compiler can separate each statement by looking at the semicolons and the curly brackets, it doesn't concern itself at all with how you format your lines. In fact, you could put the entire Blink sketch on one line if you'd like.

To make it easier for mere mortals to read and understand your code, however, it's a good idea to put each statement on its own line and indent each new block of code.

Case Sensitivity

With the Arduino language, everything is case-sensitive. This means that trying to call a function called `PINMODE()` will generate an error (try `pinMode()` instead). This goes for variables as well. If you define a variable called `led`, you can't access it with `LED`, which would be considered a totally separate variable.

Hardcoding

When you called `pinMode()` and `digitalWrite()` in the Blink sketch, you used the variable `led` to pass the pin number 13 into the function. You could have also used the functions in the following way, by putting the number 13 in each of the functions' pin parameters:

```
pinMode(13, OUTPUT);
digitalWrite(13, HIGH);
```

This is called *hardcoding* because you're writing the value directly into each function as opposed to referring to it as data from elsewhere. There are a few reasons you'll want to avoid hardcoding these values.

For one, creating a variable and giving it a good name makes your code easier to understand. If you have a motor on pin 12 and an LED on pin 13 and you want to turn the motor on and the LED off, this code is a little easier to understand:

```
digitalWrite(motorPin, HIGH);
digitalWrite(ledPin, LOW);
```

than this:

```
digitalWrite(12, HIGH);
digitalWrite(13, LOW);
```

In that last example, you'd have to remember which pin is which, or else you'd have to look it up.

Another reason to refer to the values as variables is that if you need to change your pin numbers, you won't need to dig through all the lines of your code and make changes. You'll only need to change a single line, and that will be reflected throughout the rest of your sketch.

It may seem like extra work to do it this way, but as your projects begin to grow, it will make things much easier!

Going Further with Digital Output

If you know how to control one digital pin, you know how to control the rest. Now try adding more LEDs to the mix.

For this exercise, you'll need:

- Breadboard (Makershed.com part number MKEL3, Adafruit.com part number 64, Sparkfun.com part number 12002)
- Jumper wires (Makershed.com part number MKSEEED3, Adafruit.com part number 758, Sparkfun.com part number 08431)
- LEDs (Makershed.com part number MKEE7, Adafruit.com part number 299, Sparkfun.com part number 12062)
- Resistors (Makershed.com part number MKEE4, Sparkfun.com part number 10969)

Here's how to wire up more LEDs, step-by-step:

1. Connect Galileo's ground pin to the negative rail on your breadboard.
2. Insert a few LEDs between the ground rail and different rows on the breadboard. Don't forget that the shorter lead (the cathode) connects to ground while the longer lead (the anode) connects to the digital pin through a resistor.
3. Connect each of those rows to different digital output pins through the appropriate current limiting resistor. See figure Figure 3-10.

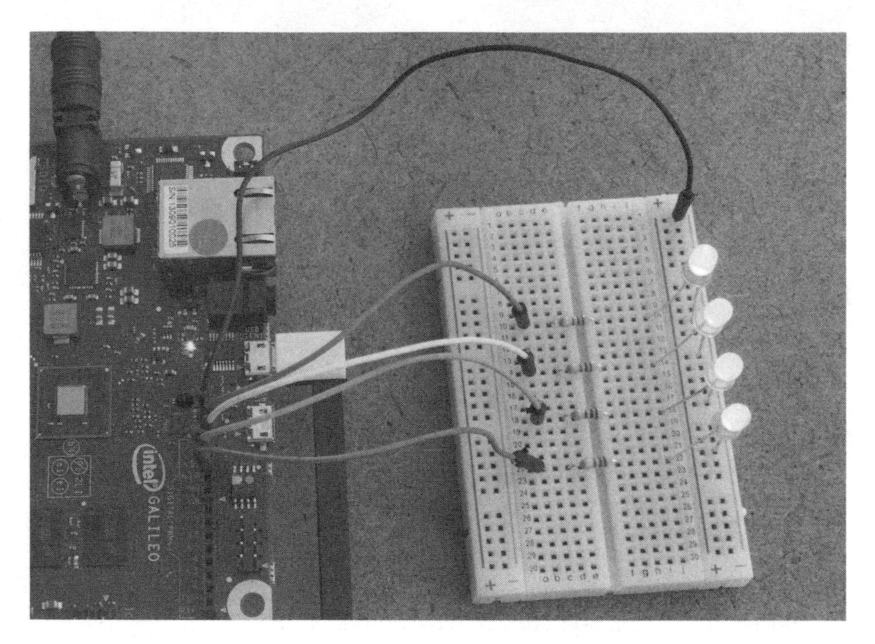

Figure 3-10. *One way of wiring up multiple LEDs to Galileo using a solderless breadboard*

Now using the functions `digitalWrite()` and `delay()`, try modifying the Blink example and uploading the code to blink the LEDs in different patterns.

Pins 0 and 1 are special digital pins because they're given additional functionality: serial communication. You'll get to try out serial in "Serial Data Output" on page 50. Be aware that if you use these pins, they may not behave as expected. For now, I would suggest avoiding digital pins 0 and 1.

Analog Output

Until now, you've been using `digitalWrite()` to control the pins. In the realm of digital, you're working with two possible states: high or low (on or off). However, not everything in the world is either on or off, and sometimes things come in a range of values.

For instance, a lamp connected to a regular wall switch is either on or off. But if it's connected to a dimmer switch (Figure 3-11), it will have a range from totally off to maximum brightness and all the levels of dim in between. If `digitalWrite()` is like an on/off switch, then the next function you'll try, `analogWrite()`, is sort of like a dimmer switch.

Figure 3-11. *Digital is like the switch on the left, it can be either on or off. Analog, on the other hand, can be set at a range of values between fully on and completely off.*

analogWrite()

I wrote that it's "sort of" like a dimmer switch because `analogWrite()` uses a feature called *pulse width modulation*, or PWM, to make it *seem* like there's a range of voltages coming out of the pin. What it's actually doing is pulsing the pins on and off really quickly. So if you want the pin to be as though it's at half voltage, the pin will be pulsed so that it is on 50% of the time and off for 50% of the time. If you want the pin to be as though it's at 20% power, it will turn the pin on 20% of the time and off 80% of the time. This percentage of time that it's on versus total time of the cycle is called the *duty cycle* (Figure 3-12). When you connect an LED to these pins and use `analogWrite()` to change the duty cycle, it can give the effect of dimming the LED.

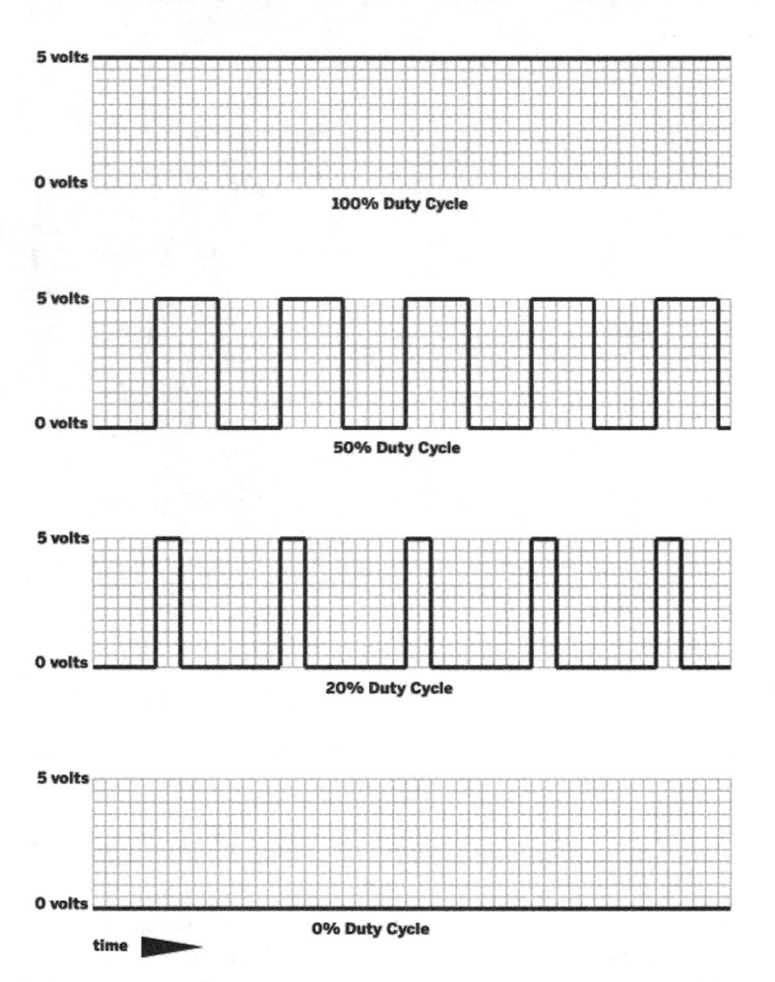

Figure 3-12. *The duty cycle represents how much time the pin is turned on over the course of an entire on-off cycle.*

However, not all pins on the Galileo are capable of being pulsed with PWM. If you look at the board, you'll see a few digital pins marked with a tilde (~), as shown in Figure 3-13. These are the pins that you can use with `analogWrite()`. On the Galileo, these are digital pins 3, 5, 6, 9, 10, and 11.

Now let's take a look at the `analogWrite()` function itself and then try to put it to use.

Figure 3-13. *You can only use pins capable of Pulse Width Modulation with* `analogWrite()`*. These pins are indicated with a tilde (~) next to the pin number on the board.*

The syntax of `analogWrite()` is:

```
analogWrite(pin, value);
```

The parameters of `analogWrite()` are:

- `pin`: the pin number
- `value`: an integer between 0 (totally off) and 255 (totally on)

`analogWrite()` does not return a value.

Try the function out now to make an LED fade up and down, much like the LED on some computers that shows it's "sleeping."

1. Connect an LED to pin 9 of your Galileo the same way you did in "Pin Numbers" on page 28. Don't forget to include a current limiting resistor in your circuit as shown in Figure 3-5.
2. Within the Arduino IDE, click File→Examples→01.Basics→Fade to open the `analogWrite()` example.
3. Click upload.

If you've got everything right, you should see a fading LED! Just like those fancy computers! Let's take a look at the code in Example 3-2.

Example 3-2. The Arduino fade example

```
int led = 9; // ❶
int brightness = 0; // ❷
int fadeAmount = 5; // ❸

void setup()  { // ❹
  pinMode(led, OUTPUT); // ❺
}

void loop()  { // ❻
  analogWrite(led, brightness); // ❼

  brightness = brightness + fadeAmount; // ❽

  if (brightness == 0 || brightness == 255) { // ❾
    fadeAmount = -fadeAmount; // ❿
  }

  delay(30); // ⓫
}
```

❶ Store the integer 9 into a new variable called `led`. This is the pin number connected to the LED.

❷ Store the integer 0 into a new variable called `brightness`. This is where the sketch will keep track of the brightness level.

❸ Store the integer 5 into a new variable called `fadeAmount`. This will define the amount of steps in brightness to take each time the brightness is changed.

❹ Start the `setup` function, which is only executed once at the start of the sketch.

❺ Set `led` (pin 9) as an output.

❻ Start the `loop` function, which is executed over and over again after the `setup` function is finished.

❼ Set `led` (pin 9) to the PWM value as determined by the variable `brightness`. The first time executing this loop, it will be 0 (completely off) because the value of `brightness` was initialized as 0 at the beginning of the sketch.

❽ Take the current value of `brightness`, add `fadeAmount` to it and then assign that amount to `brightness`. In other words, add `fadeAmount` to `brightness`.

❾ If `brightness` equals 0 or 255, execute the code within the curly brackets.

⑩ Set `fadeAmount` to its opposite (make it negative if it's positive; make it positive if it's negative).

⑪ Wait 30 milliseconds before executing the `loop` function again.

Code and Syntax Notes

Besides `analogWrite()`, the code in Example 3-2 introduces a few concepts that you haven't already encountered in this book.

Variable Assignment

Up until now, you've stored values inside variables at the time that you created the variable. In Example 3-2, you'll see that in order to assign a new value, you'll use the equal sign:

```
brightness = brightness + fadeAmount;
```

Galileo will look at the right side of the equal sign first, complete any operations, and then assign that value to the variable on the left side of the equal sign. In this case, the initial value of `brightness` is added to the value of `fadeAmount`. The result of that addition operation is then assigned as the new value of `brightness`.

When I see an equal sign used in this way, I don't think of it as "equals" but rather as "gets the value of." When reading that line in the example, I would think of it as "`brightness` gets the value of `brightness` plus `fadeAmount`." This makes it a little easier to understand what's happening in that statement.

The purpose of this line in Example 3-2 is to change the brightness every time the loop is executed.

Another variable assignment that happens in the `loop` function is:

```
fadeAmount = -fadeAmount;
```

Simply put, this sets `fadeAmount` to its opposite. If it's a negative number, it will make it positive. If it's a positive number, it will make it negative. When the `fadeAmount` is positive, it's adding to the PWM value in every loop, making it get brighter. Conversely, when it's negative, it's dimming the LED because summing the `brightness` and -5 will decrease the brightness by 5 steps.

Math Operators

Of course, you're not limited to only adding values with Arduino. You can also do subtraction (-), multiplication (*), division (/), and modulus (%), which gives the remainder when dividing two numbers.

You can use those math operations with an assignment operator to make a math operation and update the value of a variable in one fell swoop using compound operators. These two lines of code all do the exact same thing:

```
doubleThis = doubleThis * 2;
doubleThis *= 2;
```

There's also shorthand syntax to increment and decrement integers by using ++ or --. The following three lines of code all do the exact same thing:

```
countUp = countUp + 1;
countUp += 1;
countUp++;
```

And these lines of code all do the same thing as well:

```
countDown = countDown - 1;
countDown -= 1;
countDown--;
```

if Statements

Another new programming concept introduced in Example 3-2 is the `if` statement, which you'll likely find yourself using a lot. The concept is fairly simple: if something is true, then do something.

The syntax looks like this:

```
if (condition) {
    execute this code if condition is true
}
```

There are a few different conditions you can test for, as laid out in Table 3-1.

Table 3-1. *Comparison operators*

Operator	Evaluates to True if...
x == y	x is equal to y
x != y	x is not equal to y
x < y	x is less than y
x > y	x is greater than y
x <= y	x is less than or equal to y
x >= y	x is greater than or equal to y

Let's take a look at a few simple examples of `if` statements:

```
int n = 10;

if (n > 10) {
```

```
    // this will not be executed because n is not greater than 10
    digitalWrite(redLed, HIGH);
}

if (n < 10) {
    // this will not be executed because n is not less than 10
    digitalWrite(greenLed, HIGH);
}

if (n == 10) {
    // this will be executed because n equals 10
    digitalWrite(yellowLed, HIGH);
}
```

It's important to remember that a single equal sign is used to assign values to variables and a double equal sign is a comparison operator to check if one value is the same as another. It's very unlikely that you'll ever use the assignment operator inside an `if` statement's parentheses. I find that it's an easy mistake to make. Unfortunately, you will not get an error from the compiler if you make this mistake, and your sketch won't behave as expected.

You can have multiple tests in a single `if` statement. The *logical operators* to help you do this are listed in Table 3-2. There's also a *not* operator to negate any result.

Table 3-2. *Logical operators*

Operator	Meaning
&&	and
||	or
!	not

Here are a few examples to demonstrate how the logical operators work so that you can test for two conditions at once:

```
int n = 10;

if ( (n > 8) && (n < 12) ) {
    // this will be executed because n is greater than 8 and less
    // than 12
    digitalWrite(redLed, HIGH);
}

if ( (n > 8) && (n < 10) ) {
    // this will not be executed because n is not less
    // than 10 even though it is greater than 8. With &&,
    // both must be true.
```

```
        digitalWrite(greenLed, HIGH);
    }

    if ( (n > 8) || (n < 10) ) {
        // this will be executed because n is greater than 8, even though
        // it's not less than 10. With ||, only one must be true.
        digitalWrite(greenLed, HIGH);
    }
```

With that information, you can now better understand what's going on in the `if` statement in Example 3-2:

```
if (brightness == 0 || brightness == 255) {
  fadeAmount = -fadeAmount ;
}
```

If the value of `brightness` equals 0 or it equals 255, then set the value of `fadeAmount` to its opposite. Essentially, when the brightness hits the maximum value (255), the sketch starts fading the LED down until it gets to its minimum value (0). Then it reverses `fadeAmount`'s direction again.

Other Outputs

Output pins aren't just for blinking and fading LEDs. They can also be used to do things like control motors, make sounds, or communicate with devices.

Serial Data Output

Sometimes you may want to have Galileo send data to your computer. Perhaps you need help figuring out why a project isn't working (also called *debugging*). Or you want to send sensor data to a spreadsheet. Maybe you want to use the Galileo as a controller for a computer game you made.

For these purposes, you can use Arduino's serial library, which is a way of sending and receiving data between devices. For now, you're just going to use Galileo to send serial data, but you'll learn how to receive data as well in Chapter 5.

The USB connection between your computer and Galileo is not only for programming the board. It's also for serial communication. To experiment with basic serial communication from Galileo to your computer, make a few simple modifications to the fade sketch, Example 3-2. These changes are reflected in Example 3-3.

Example 3-3. The Arduino fade example with serial

```
int led = 9;
int brightness = 0;
int fadeAmount = 5;

void setup()  {
  pinMode(led, OUTPUT);
  Serial.begin(9600); // ❶
}

void loop()  {
  analogWrite(led, brightness);

  brightness = brightness + fadeAmount;

  if (brightness == 0 || brightness == 255) {
    Serial.print ("Brightness is at "); // ❷
    Serial.print (brightness); // ❸
    Serial.println (". Switching directions."); // ❹
    fadeAmount = -fadeAmount;
  }

  delay(30);
}
```

❶ Opens serial port and sets data rate to 9600 bits per second (also known as the baud rate).

❷ Send a string of text over serial.

❸ Send the value of brightness over serial.

❹ Send a string of text over serial and then a carriage return.

After uploading the code from Example 3-3 to the board, click the magnifying glass button on the right side of the Arduino IDE toolbar to open the serial monitor (as shown in Figure 3-14).

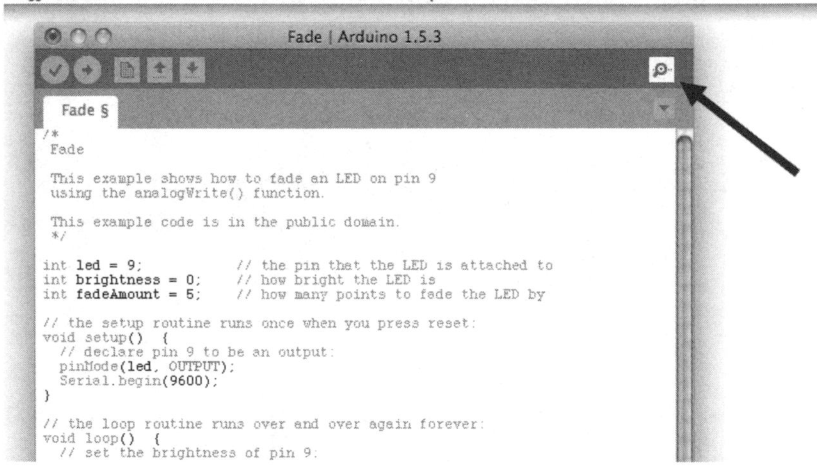

Figure 3-14. *To open the serial monitor, click the magnifying glass button on the upper-right side of the Arduino IDE window.*

A new window will appear. Make sure that 9600 baud is selected in the drop-down on the lower-right side of the window. If you've got everything right, you should see something like this, over and over again (see Figure 3-15):

```
Brightness is at 0. Switching directions.
Brightness is at 255. Switching directions.
Brightness is at 0. Switching directions.
Brightness is at 255. Switching directions.
Brightness is at 0. Switching directions.
Brightness is at 255. Switching directions.
```

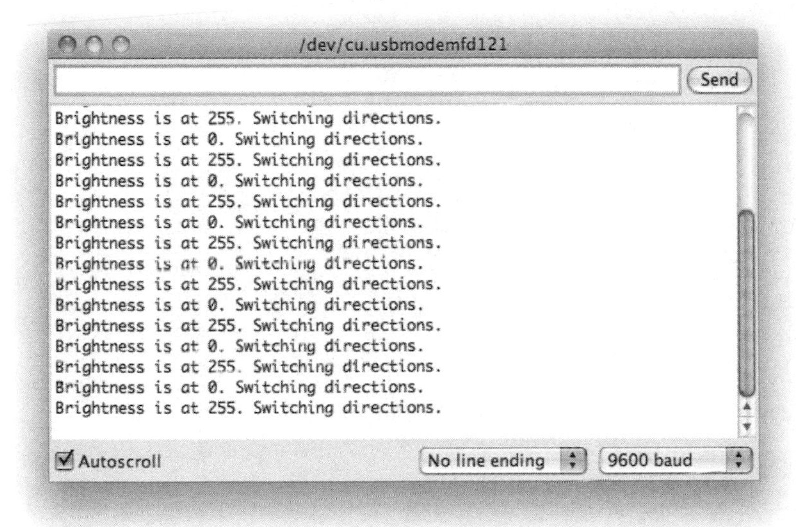

Figure 3-15. *Set the serial monitor's baud rate to match the baud rate established by the `Serial.begin()` function in your sketch.*

To get this to happen, you're using three different serial library functions. Let's look at each.

Serial.begin()

This is the function that opens the serial port on the Galileo and tells it what speed to send the data. The number represents bits per second. Because this is only used once in a sketch, you'll typically call it in the `setup` function.

Two devices communicating via serial must be using the same baud rate, even if you're only transmitting data in one direction. If you have a mismatch between the baud rate of your two devices, the receiving device may interpret the data it receives as gibberish.

However, as a side effect of how Galileo handles serial communication between the board and the serial monitor, setting matching baud rates may not matter. It would be a good idea to do it anyway so that you're accustomed to doing it and so that your code will work with other Arduino boards.

Within this book, I'll be sticking with standard Arduino convention because it will work with Galileo and other Arduino boards.

The syntax of `Serial.begin()` is:

```
Serial.begin(speed);
```

The parameter is:

`speed`: the speed in bits per second (also known as baud). The following values are typically used: 300, 600, 1200, 2400, 4800, 9600, 14400, 19200, 28800, 38400, 57600, or 115200. With Arduino sketches, 9600 is most common.

`Serial.begin()` does not return a value.

Serial.print()

This function transmits data over the serial port.

The syntax of `Serial.print()` is:

```
Serial.print(value);
```

The parameter is:

`value`: the data to send. This can be a string of text, a character, a byte, an integer, or other types of data.

`Serial.print()` returns the number of bytes transmitted. It's not necessary to use this return value and you can safely ignore it.

Serial.println()

Just like `Serial.print()`, `Serial.println()` sends data over the serial port. The difference is that it adds a carriage return to the end of the data it sends.

The syntax of `Serial.println()` is:

```
Serial.println(value);
```

The parameter is:

`value`: the data to send. This can be a string of text, a character, a byte, an integer, or other types of data.

`Serial.println()` returns the number of bytes transmitted. It's not necessary to use this return value and you can safely ignore it.

Controlling A/C Appliances with Relays

The small amount of electrical current that comes from the digital pins is not enough to power much more than a small LED. In order to control the power to something like a lamp or blender, you'll need to use a *relay*.

A typical relay is a mechanical switch that can handle larger amounts of electricity. That switch can be turned on or off with a small amount of electricity.

It's possible to buy a relay and then do the wiring so that it's switching the power source that you want to control. However there's a fantastic product called the PowerSwitch Tail which makes it easy to use a development board like the Galileo to control appliances that plug into the wall. You don't need to do much wiring at all because your appliance will plug right into the PowerSwitch Tail and the PowerSwitch Tail will plug into your wall outlet (see Figure 3-16).

Figure 3-16. *The PowerSwitch Tail helps you move from controlling LEDs to controlling high voltage A/C devices that plug into your wall outlet.*

You only need to wire up ground and one of Galileo's digital out pins to the Power-Switch Tail, then it's only a matter of using `digitalWrite` to turn it on and off:

```
// Turn the blender on for one second:
digitalWrite(powerSwitchTailPin, HIGH);
delay(1000);
digitalWrite(powerSwitchTailPin, LOW);
```

Controlling Servos

A typical hobby servo motor is a small motor that can set and hold the position of its axle depending on the pulses of electricity being sent to it, usually with a 180 degrees of rotation (Figure 3-17). It's powered by a connection to

5 volts and ground through the servo's red and black wires. Your Galileo will send signal pulses to the motor to set its position through its yellow wire.

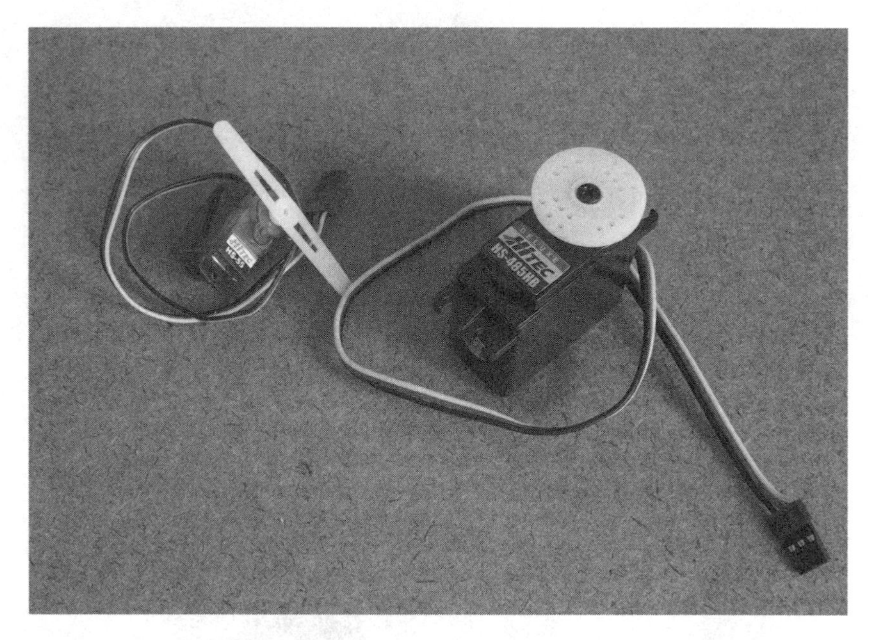

Figure 3-17. *Hobby servos like the ones pictured come in many different sizes and varieties but generally operate in the same way.*

Sending the right high and low pulses to a servo (or servos) might be a bit difficult and time consuming if you had to write the code by yourself. Luckily, you have access to an Arduino library to help you work with servo motors with very little fuss. A *library* is a collection of code that makes complex coding tasks easier. Instead of needing to know what rate you need to send pulses and how to send them without blocking the rest of your code from running, you can simply input the angle into a function from the library. The Galileo will start sending the appropriate pulse until you tell it to stop or change the angle (Figure 3-18).

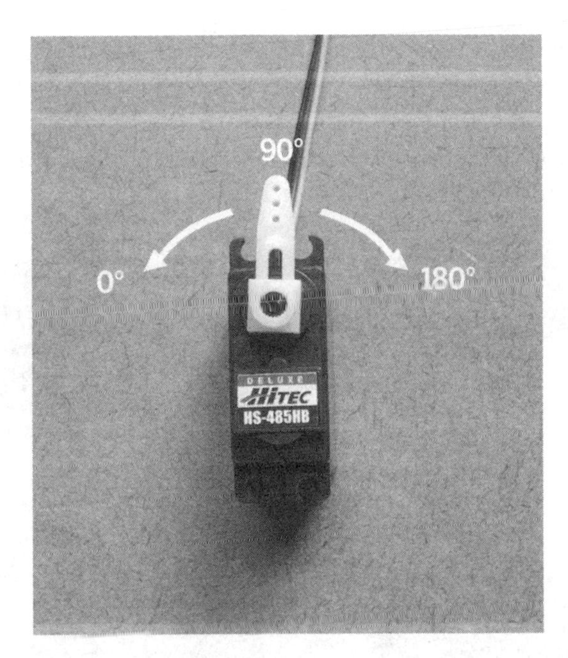

Figure 3-18. *Most servos have 180 degrees of rotation. When using the Arduino servo library, 0 degrees positions the axle in the fully counterclockwise position. 180 degrees positions the axle in the fully clockwise position. 90 degrees sets the axle in the middle.*

To start experimenting with servos and try the servo library, first connect a servo to your Galileo.

For this exercise, you'll need:

- Analog hobby servo (There are many types, but any of the following should work: Makershed.com part number MKMSERVO, Adafruit.com part number 169, Sparkfun.com part number 09065)
- Jumper wires (Makershed.com part number MKSEEED3, Adafruit.com part number 758, Sparkfun.com part number 08431)

Wiring up and testing the servo won't be difficult:

1. Using a jumper wire, connect the black wire from the servo to ground on the Galileo. You can connect the jumper wire directly from the Galileo to the servo wires, or you can use a breadboard. You'll need to insert male header pins into the servo plug so that they can be inserted into a breadboard.
2. Connect the red wire from the servo to the VIN pin on the Galileo. This pin is connected to the positive voltage of your power source.

3. Connect the yellow wire from the servo to digital pin 9. It should look something like Figure 3-19 or Figure 3-20.
4. In the Arduino IDE, enter the code from Example 3-4 into a new sketch.
5. With the Galileo powered on and connected to your computer via USB, upload the sketch to the board.

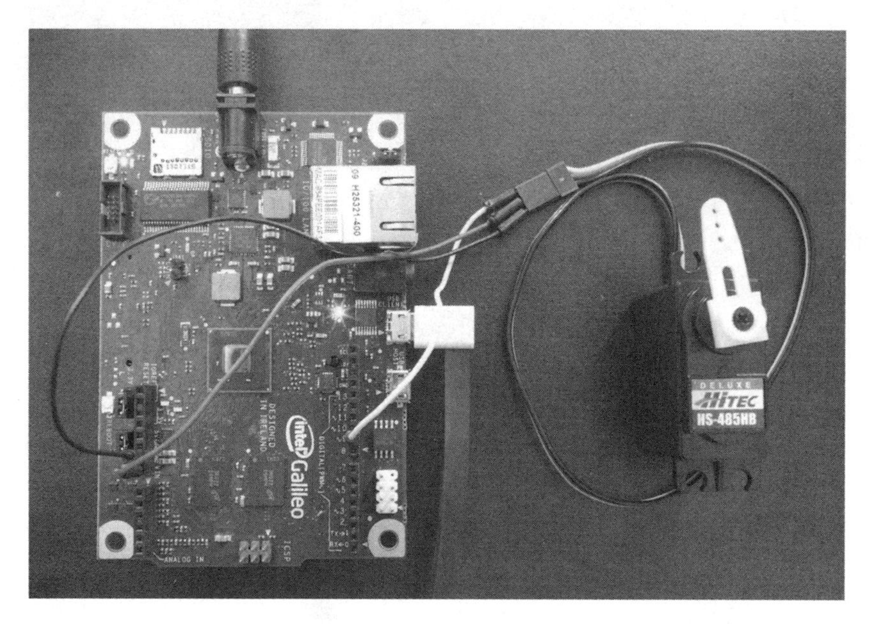

Figure 3-19. *Wiring up a hobby servo to Galileo*

If you have everything wired up correctly, you should see the servo swing back and forth (see Example 3-4)!

Example 3-4. Servo Test

```
#include <Wire.h>
#include <Servo.h> // ❶

int servoPin = 9;

Servo myServo; // ❷

void setup() {
  myServo.attach(servoPin); // ❸
}

void loop() {
  myServo.write(0); // ❹
  delay(1000);
```

```
  myServo.write(180); // ❺
  delay(1000);
}
```

❶ When compiling, include the servo library code. This is required to use the servo library's functions.

❷ Create a new servo object called `myServo`.

❸ Establish that the servo you'll be controlling is on pin 9.

❹ Set the servo's position to 0 degrees.

❺ Set the servo's position to 180 degrees.

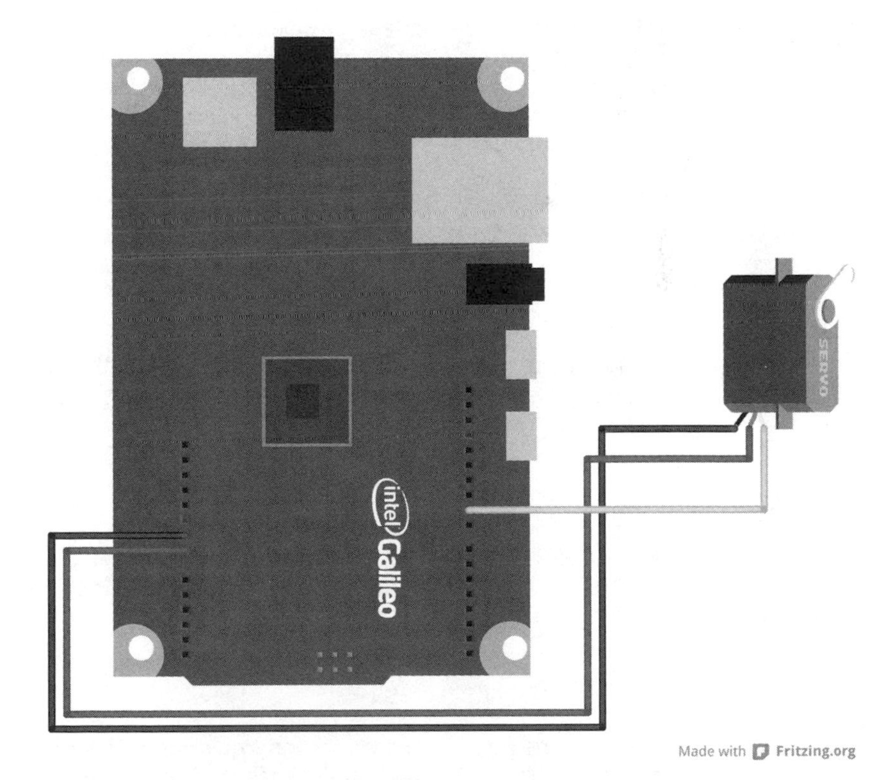

Figure 3-20. *The servo's yellow wire connects to your digital output pin (any of the PWM pins will work). The black (or sometimes brown) wire connects to ground. The red wire connects to the VIN pin.*

In Example 3-4, the line `#include <Wire.h>` is a workaround to get the servo library working with Galileo in Intel's version 0.7.5 of the software. It may not be required for later versions. Your best bet is to try it with and without the line to see what works.

The Servo Object

When you're reviewing the lines of code that control the servo, you're encountering a new programming concept called the *object*.

```
Servo myservo;
```

In programming parlance, this line creates a `Servo` object called `myservo`. You don't need to know the nitty gritty of what an object is or how it works, but what's important with an object is that it lets you create multiple servos and act on them independently with the servo functions. The best way to see this in action is to look at an example with two servos. Let's say you have a camera on a servo that controls its pan from left to right and another servo that controls the camera's tilt from up to down. To do this, you would create two `Servo` objects:

```
Servo panServo;
Servo tiltServo;
```

You can then tell Galileo which pin each servo is connected to with the `attach()` function. Assuming the pan servo is on pin 9 and the tilt servo is on pin 10, you'd put the following in your setup function:

```
panServo.attach(9);
tiltServo.attach(10);
```

Then to tell each servo where to go, you'll use the `write()` function, and input the number of degrees between 0 and 180. If you wanted each servo to sit right in the middle, it would be 90 degrees:

```
panServo.write(90);
tiltServo.write(90);
```

Those are the basics of how the servo library works and a quick crash course in creating and acting on an object. Most other libraries that you'll use will use the object paradigm and usually include examples to help you understand them.

Looking at Linux

Up until this point in the book, most of what you've learned has focused on the Arduino-like capabilities of Galileo. As mentioned in Chapter 1, one of the features of Galileo is that it's running Linux. Let's now take a small dip into the world of Linux on Galileo.

The first thing you need to do is connect to the Linux command line. Here, I'll show you how to connect via Telnet over the network. You can also connect directly via serial. See Appendix H for how to do that.

Connecting via Telnet

If your computer is connected to the same LAN (local area network) as the Galileo, you can use Telnet to connect to Galileo's Linux command line prompt through the network.

1. Using an Ethernet cable, connect your Galileo to the same network as your computer.
2. With your Galileo powered up and connected to your computer via USB, upload Example 3-5 to the board. This code will enable Telnet on the board and pipe information about Galileo's network connection to the serial monitor.

Example 3-5. Code to enable Telnet and Print IP

```
void setup() {
  system("telnetd -l /bin/sh"); // ❶
}

void loop() {
  system("ifconfig eth0 > /dev/ttyGS0"); // ❷
  delay(5000);
}
```

❶ Execute the Linux command to enable Telnet.

❷ Output information about the Ethernet connection to the Arduino serial monitor.

Example 3-5 uses the `system()` function, which is unique to Galileo. It's meant for running Linux commands within Arduino code. This will be covered in more depth in Chapter 6.

3. Open the serial monitor and get the IP address, indicated by `inet addr` in the response. If you don't see `inet addr`, make sure your board is connected to the router and try rebooting it and going through these steps again.

```
eth0      Link encap:Ethernet  HWaddr 98:4F:EE:00:1A:F3
          inet addr:192.168.1.4  Bcast:0.0.0.0  Mask:255.255.255.0
          inet6 addr: fe80::9a4f:eeff:fe00:1af3/64 Scope:Link
          UP BROADCAST RUNNING MULTICAST  MTU:1500  Metric:1
          RX packets:326 errors:0 dropped:0 overruns:0 frame:0
          TX packets:93 errors:0 dropped:0 overruns:0 carrier:0
          collisions:0 txqueuelen:1000
          RX bytes:52925 (51.6 KiB)  TX bytes:7511 (7.3 KiB)
          Interrupt:41 Base address:0x8000
```

Next, connect to your Galileo:

On Mac OS X

1. Launch Terminal from */Applications/Utilities*.
2. From your computer's command line, type `telnet 192.168.1.4` (replacing `192.168.1.4` with your Galileo's IP address).

On Linux

1. Launch Terminal by typing Control+Alt+T or finding the application within your launcher.
2. From your computer's command line, type `telnet 192.168.1.4` (replacing `192.168.1.4` with your Galileo's IP address).

On Windows

1. Click Start→Run.
2. In the text box, type `telnet 192.168.1.4` (replacing `192.168.1.4` with your Galileo's IP address).

Once you're connected, you'll see a # to indicate you're at the command prompt:

```
Trying 192.168.1.4...
Connected to 192.168.1.4.
Escape character is '^]'.

Poky 9.0 (Yocto Project 1.4 Reference Distro) 1.4.1 clanton

/ #
```

Any changes you make to files will not persist after a reboot unless you're booting from a microSD card. See Appendix D for how to create one.

Working with Pins

From the command-line prompt, you can execute Linux commands on Galileo. You can use commands to read or write files, get information about your system, make network connections, read and write the pins, and much more. You're going to use a few Linux commands to turn a pin into an output and then set it to high.

From the command line, first log in as root (you don't need to do this if you're connected via serial):

```
/ # login root
root@clanton:~#
```

Next use the command `cd` to change to the directory `/sys/class/gpio`:

```
root@clanton:~# cd /sys/class/gpio/
```

Now list the contents of that directory with the command `ls`:

```
root@clanton:/sys/class/gpio# ls
export    gpio19    gpio27    gpio38    gpio46    gpiochip0
gpio0     gpio20    gpio28    gpio4     gpio47    gpiochip16
gpio1     gpio21    gpio29    gpio40    gpio48    gpiochip2
gpio14    gpio22    gpio30    gpio41    gpio49    gpiochip8
gpio15    gpio23    gpio31    gpio42    gpio50    unexport
gpio16    gpio24    gpio32    gpio43    gpio51
gpio17    gpio25    gpio36    gpio44    gpio54
gpio18    gpio26    gpio37    gpio45    gpio55
```

This is a listing of files and *directories* (or folders) for working with the different pins on Galileo (it's been formatted to fit this page nicely, so it may not look exactly like your output). Within the Linux environment, you can write to a particular file to control a pin and read from a file to get the pin's state.

The letters GPIO above stand for *general purpose input/output*, which means that the pins can be configured to do many different things such as digital output, digital input, analog output, and analog input, to name a few.

On other Linux development boards, you normally would not necessarily see all the GPIO pins in `/sys/class/gpio`. However, since you've uploaded and launched an Arduino sketch to get Telnet running, the sketch *exported* those pins and configured them for use already.

The listing of the pin numbers in `/sys/class/gpio` does not match the pin numbers you use within your Arduino code or the numbers printed on the board. Table 3-3 shows which Arduino pin number matches which Linux GPIO signal number.

Table 3-3. *Translating Arduino pin numbers to Linux signal names*

Arduino Digital Pin	Linux Signal Number
0	50
1	51
2	14
3	15
4	28
5	17
6	24
7	27
8	26
9	19
10	16
11	25
12	38
13	39

Connect an LED to pin 13, just as you did in "Pin Numbers" on page 28. Because pin 13 is connected to Linux signal name 39 as shown in Table 3-3, change to the directory `gpio39` and list its contents.

```
root@clanton:/sys/class/gpio# cd gpio39
root@clanton:/sys/devices/virtual/gpio/gpio39# ls
active_low  direction   edge        power       subsystem   uevent
value
```

The next command you'll use is `cat`, which outputs the contents of a file to the terminal. First output the contents of the `direction` file.

```
root@clanton:/sys/devices/virtual/gpio/gpio39# cat direction
in
```

This indicates that the pin is configured as an input. Change it to output by using the command `echo` to write "out" to gpio39's `direction` file. This is the equivalent of the Arduino code `pinMode(13, OUTPUT);`:

```
root@clanton:/sys/devices/virtual/gpio/gpio39# echo out > direction
```

To set pin 13 high, just write the value 1 to the `value` file. This is the equivalent of the Arduino code `digitalWrite(13, HIGH);`:

```
root@clanton:/sys/devices/virtual/gpio/gpio39# echo 1 > value
```

If you got it right, the LED should have turned on! Now turn it off:

```
root@clanton:/sys/devices/virtual/gpio/gpio39# echo 0 > value
```

When you toggled pin 13 with Arduino code, you also controlled the on-board LED pictured in Figure 2-8. Why didn't it turn on and off now? That's because the on-board LED on Galileo is connected to Arduino pin 13 through software only. It's actually on its own Linux GPIO signal. If you want to try blinking it, it's controlled with Linux signal name gpio3.

Because you can read and control Galileo's pins by reading and writing files within the Linux environment, it opens up a whole new realm of power and flexibility, one that previous Arduino boards didn't have. As the Galileo software matures, it's likely that it will take advantage of this ability and further empower users. For now, consider yourself in somewhat uncharted territory and report back what you find.

Taking It Further

This chapter armed you with the ability to control digital output pins by writing their values to be high or low or using pulse width modulation to send pulses of high and low signals. Along the way, you learned a lot of programming concepts that you'll use throughout your work with Galileo.

Here are a few ideas for how you can apply what you've learned in this chapter:

- Create an LED light show timed to sync up with a song using `digitalWrite()`, `analogWrite()`, and `delay()`.
- Create a sketch that pulses an LED to communicate a message using Morse code.
- Strap a small hourglass to a servo's arm and see if you can program the timing on the Galileo so that it flips the hourglass over immediately after all the sand falls through.

Here are some additional resources in case you'd like to learn more about the concepts covered in this chapter:

- There's an example in the Arduino IDE that shows you how to blink an LED without using the `delay()` function. This will come in handy to learn how to set something to run on an interval without blocking other operations, which is what `delay()` does. To try it out, click File→Examples→02.Digital→BlinkWithoutDelay.
- My friend Collin made a great video explaining pulse width modulation (*http://www.youtube.com/watch?v=Lf7JJAAZxEU*).

- The data sent via serial is in the ASCII character encoding scheme and the byte values for each character can be represented in a few different ways. To explore this more, I recommend reviewing the ASCIITable example. Just click File→Examples→04.Communication→ASCIITable.

4/Inputs

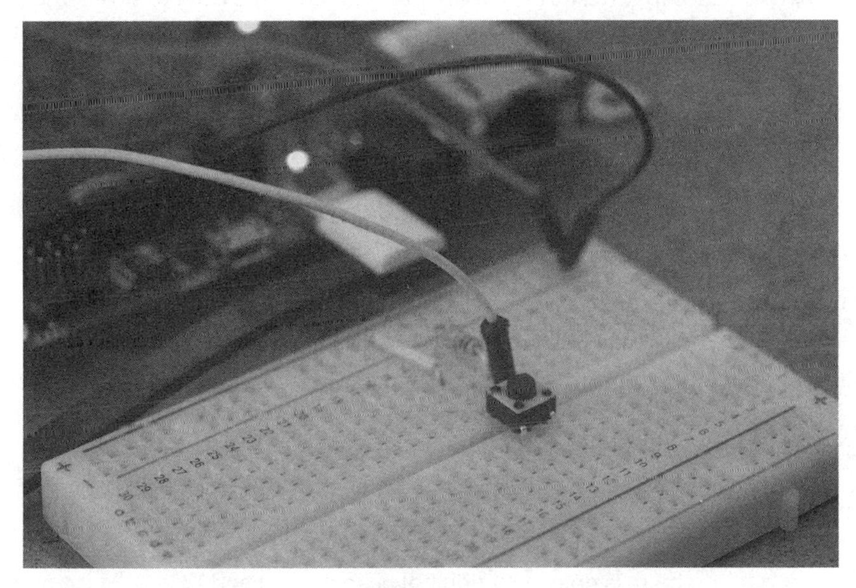

Buttons, switches, dials, cameras, motion sensors, and pressure sensors are all examples of possible inputs to Galileo, and that list only scratches the surface of inputs available to you. Inputs help Galileo know what's happening in the physical world so that a user can control your device or the device can respond to its environment. When combined with outputs, inputs make your device interactive.

In this chapter, you will use Galileo's input functionality to:

- Understand how to read digital and analog inputs
- Experiment with a few different types of inputs
- Learn more about coding with Arduino

- Learn how inputs and outputs can work together to make something interactive

Switches: Digital Input

When you used `digitalWrite()` in Chapter 3, there were two possible states of a digital output pin: high or low. When it comes to digital input, the same is true. When you read the state of a digital input, it will either be connected to 5 volts, which indicates a *high* state, or to ground, which indicates a *low* state. By connecting a simple pushbutton switch, you can change the state of a digital input pin simply by pressing the button.

A *pushbutton switch* like the one pictured in Figure 4-1 is great for experimenting with digital input. It snaps right into the breadboard, and it's a component that you'll frequently see in electronics starter kits. The two terminals on the top of the switch are connected to each other, as are the pair on the bottom. When you push down on the switch, both pairs of terminals are connected together.

Because the terminals on opposite sides of the breadboard are connected to each other, it means that the pushbutton is making a connection across the gap in the breadboard just as if you had put a jumper wire across the gap. So, in Figure 4-1, this means that both sides of row 11 are connected, as are both sides of row 13. But unless you are pressing the button down, there is no connection between rows 11 and 13.

Figure 4-1. *A pushbutton switch*

You're going to use a pushbutton switch to make and break a connection between 5 volts and one of the digital pins, which will be configured in your sketch as an input. You're also going to add a *pulldown resistor* to ensure that the input is connected to ground when it's not connected to 5 volts. If the input pin were disconnected from both 5 volts and ground, it would be considered *floating* and it would return unexpected results if you tried to read it. That is because an unconnected pin can be affected by ambient electrical activity (without the pulldown resistor, the pin essentially functions as an antenna). See Figure 4-4 for a closer look at the pulldown resistor in the circuit.

Here's what you'll need to try this example:

- Breadboard (Makershed.com part number MKEL3, Adafruit.com part number 64, Sparkfun.com part number 12002)
- Jumper wires (Makershed.com part number MKSEEED3, Adafruit.com part number 758, Sparkfun.com part number 08431)
- Pushbutton switch (For the most basic type of switch that will fit in a breadboard: Adafruit.com part number 00097, Sparkfun.com part number 00097)
- 10K ohm resistor (Assorted resistor packs: Makershed.com part number MKEE4, Sparkfun.com part number 10969)

Here's how to wire it up and try it out:

1. Insert your pushbutton switch into the breadboard as shown in Figure 4-1. If it doesn't snap in easily, it may not be oriented correctly, so try rotating the switch 90 degrees.
2. Using a jumper wire, connect digital pin 2 of the Galileo to one of the top terminals of the switch.
3. Connect one of the bottom terminals of the switch to 5 volts.
4. Using a 10K resistor (brown, black, orange, and then gold or silver stripes), connect the top terminal of the switch to ground. Your circuit should look like Figure 4-2 or Figure 4-3.
5. Open a new sketch in the Arduino IDE and enter the code in Example 4-1.
6. Upload this sketch to Galileo.

Figure 4-2. *Connecting a pushbutton switch to digital pin 2 with a 10K pull-down resistor*

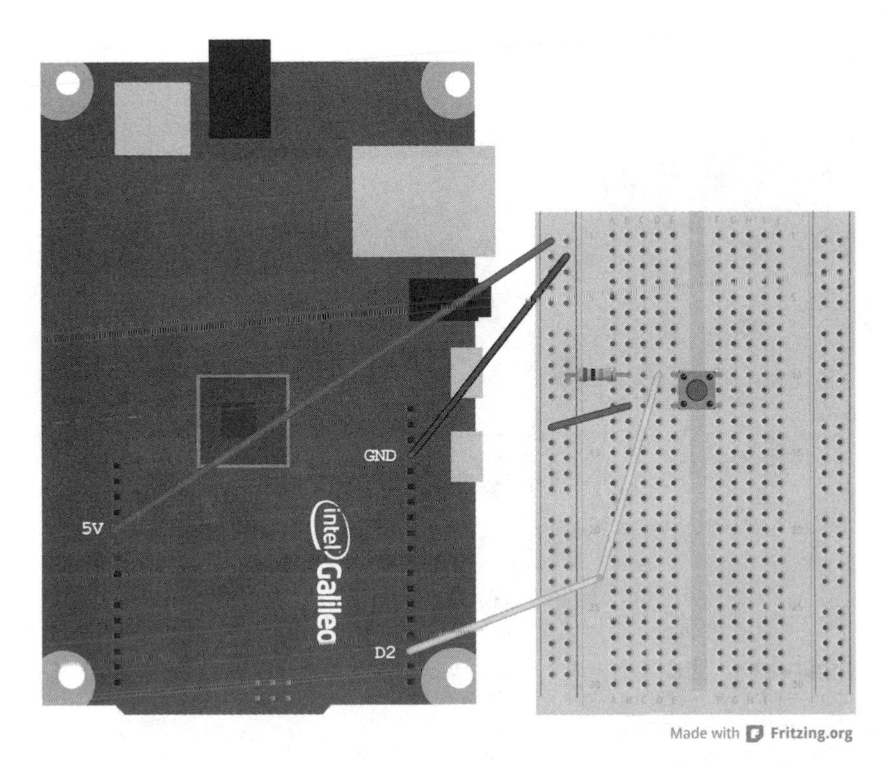

Figure 4-3. *Connecting a pushbutton switch to digital pin 2 with a 10K pull-down resistor*

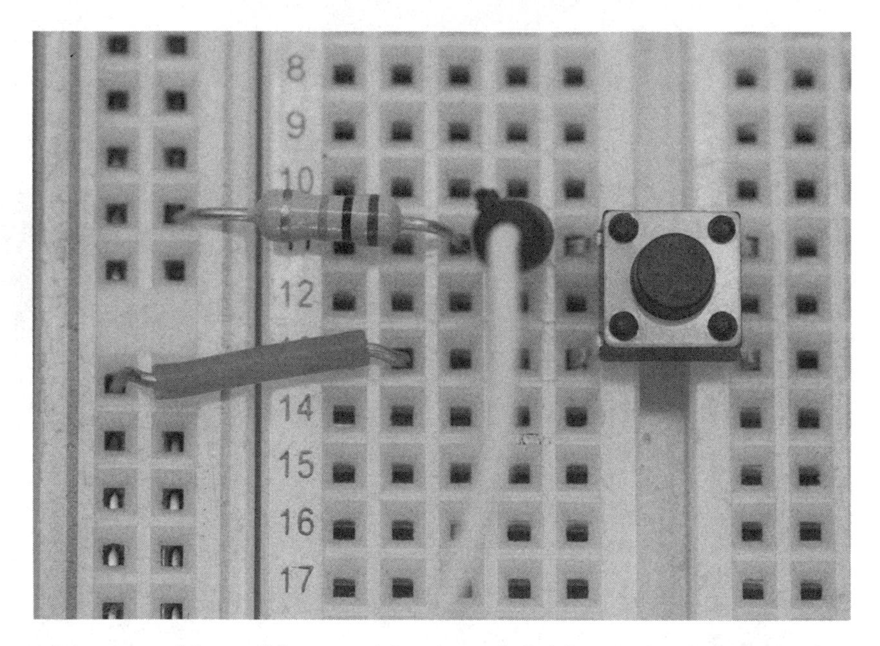

Figure 4-4. *The pulldown resistor in a digital input circuit ensures that there's a connection to ground when the connection between 5 volts and the input pin is broken.*

Example 4-1. Basic digital input sketch

```
int switchInputPin = 2;

void setup() {
        pinMode(switchInputPin, INPUT); // ❶
        Serial.begin(9600);
}

void loop() {
        int switchState = digitalRead(switchInputPin); // ❷
        if (switchState == HIGH) { // ❸
                Serial.println("The switch is on!"); // ❹
        }
        else { // ❺
                Serial.println("The switch is off!"); // ❻
        }
        delay (500); // ❼
}
```

❶ Set the `switchInputPin` (pin 2) as an input.

❷ Read the state of `switchInputPin` and store it in the variable `switchState`.

❸ If `switchState` is high or connected to 5 volts...

❹ ...print "The switch is on!" via serial.

❺ If `switchState` is low or connected to ground...

❻ ...print "The switch is off!" via serial.

❼ Pause for half a second to slow down the output of the sketch.

When you upload this code to Galileo and open the serial monitor, you should see `The switch is off!` printed repeatedly in the window. Push down on the switch and it should change!

digitalRead()

The main takeaway from Example 4-1 is the use of `digitalRead()`, which checks the value of the pin provided in the parameters. In this case, you provided the variable `switchInputPin`, which evaluates to 2. After checking to see if pin 2 is connected to 5 volts or ground, `digitalRead()` returns either HIGH or LOW respectively. That value is stored in a new variable called `switchState`.

You may be wondering: HIGH and LOW aren't integers, so why are we storing them as variables with the type `int`? The reason is that Galileo thinks of high and low as the integers 1 and 0 and therefore `HIGH` and `LOW` are defined to mean 1 and 0 respectively in Arduino code. This makes the code a little bit easier to read and understand.

To prove that `HIGH` and `LOW` are equivalent to 1 and 0, try using `Serial.println()` to print the value of `HIGH + HIGH` to the serial monitor.

You can also see that you used `pinMode()` to set pin 2 as an input in the setup function. You're required to do this if you want to use `digitalRead()` on that pin.

The syntax of `digitalRead()` is:

```
digitalRead(pin);
```

The parameter of `digitalRead` is:

- `pin`: the input pin number you want to read

`digitalRead()` returns HIGH or LOW.

Code and Syntax Notes

There are also a couple of new programming concepts introduced in Example 4-1.

Local Variables

In the code examples in Chapter 3, variables were declared outside of the `setup` and `loop` functions. Those were considered *global variables* because they can be accessed and changed from either the `setup` or `loop` functions. And when you learn to write your own functions, you could access global variables within those functions as well.

However, in Example 4-1, you declared a new variable within the `loop` function:

```
int switchState = digitalRead(switchInputPin);
```

What's important to know about variables declared within a block of code is that it can only be accessed within that block of code. It's called a *local variable*. When the Galileo is done executing that particular block, it frees up that memory for other uses. The block of code where a variable can be accessed is called its *scope*.

Therefore, when the loop function has completed a cycle, the `switchState` variable is destroyed. A new `switchState` variable is created the next time the loop function is executed.

if... else Statements

Building on what you learned in "if Statements" on page 48, the `else` statement in Example 4-1 is a way of setting up a block of code to execute when the `if` condition evaluates as false.

Here's the basic syntax:

```
if (condition) {
        execute this code if condition is true
}
else {
        execute this code if condition is false
}
```

In Example 4-1, the `if` statement checks if the input pin is high. If it is, it will execute the block of code immediately after it in order to print the text "The switch is on!" However, if the pin is low, the `if` condition will evaluate as false and then Galileo will execute the block of code immediately following the `else` statement. This will print "The switch is off!"

You'll never see an `else` statement without an `if` statement. But as you saw in Example 3-2, you can have an `if` without an `else`.

Analog Input

As discussed in Chapter 3, a digital pin (input or output) represents information as either on or off. But there are plenty of inputs that could have a range of values. Dials, sliders, temperature sensors, and light sensors are all examples of possible analog inputs to Galileo. Along with the function `analogRead()`, you can have Galileo get the value of these inputs and act on them.

On the Galileo, there are six dedicated analog input pins numbered 0 through 5 (Figure 4-5). They accept a range of voltages from 0 to 5 volts. Since Galileo and all computers work in the digital realm, the range of voltages must be converted into digital data. Therefore, these input pins connect to an *analog to digital converter*, or ADC, which is a chip that allows Galileo to read what voltage is being sent to each pin.

Figure 4-5. *The six analog input pins are numbered 0 through 5.*

Thanks to Galileo and the Arduino software, you don't need to understand how an ADC works in order to read the value of analog sensors in your projects.

Potentiometers

When first experimenting with analog input, I recommend trying out a *potentiometer*. Frequently called *pots*, these components connect to voltage and ground and deliver a variable voltage to the analog input pin. They can come in a few different varieties such as *rotary* (Figure 4-6) or *linear*, like the faders you'd see on a sound mixing board.

Figure 4-6. *This is a typical rotary potentiometer.*

Let's wire up a potentiometer to Galileo to give its ADC a test drive. Here's what you'll need:

- Breadboard (Makershed.com part number MKEL3, Adafruit.com part number 64, Sparkfun.com part number 12002)
- Jumper wires (Makershed.com part number MKSEEED3, Adafruit.com part number 758, Sparkfun.com part number 08431)
- 10k ohm potentiometer (Adafruit.com part number 562, Sparkfun.com part number 09288)

Here's how to connect the potentiometer to Galileo:

1. Connect Galileo's ground pin to the negative rail of your breadboard.
2. Connect Galileo's 5-volt pin to the positive rail of your breadboard.

3. Insert a 10K potentiometer into the breadboard.
4. The pot will have three pins. Connect the pin on one side of the potentiometer (it doesn't matter which) to the breadboard's positive rail.
5. Connect the pin on the other side of the potentiometer to the negative rail.
6. Connect the center pin of the potentiometer to analog pin 0. Your circuit should look something like Figure 4-7 or Figure 4-8.
7. In the IDE, input the code from Example 4-2 and upload it to Galileo.
8. After uploading the code and opening the serial monitor, watch the text as you turn the dial on the potentiometer from one side to another. You should see something like this, or perhaps the reverse:

```
Potentiometer is at 0%.
Potentiometer is at 5%.
Potentiometer is at 9%.
Potentiometer is at 14%.
Potentiometer is at 22%.
Potentiometer is at 30%.
Potentiometer is at 39%.
Potentiometer is at 50%.
Potentiometer is at 62%.
Potentiometer is at 74%.
Potentiometer is at 89%.
Potentiometer is at 100%.
```

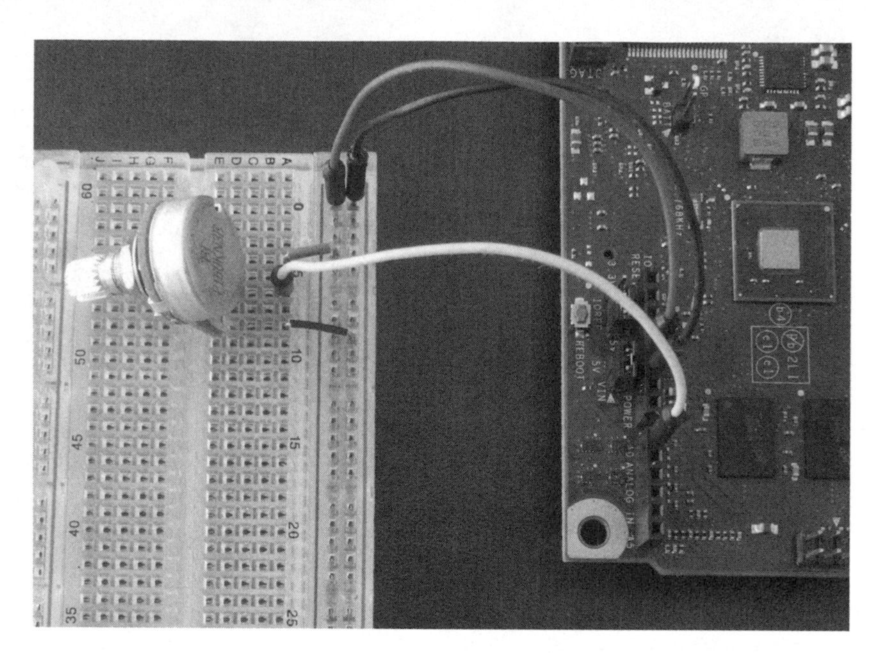

Figure 4-7. *There are three terminals on a potentiometer. The one in the middle connects to analog input pin 0. One of the outside terminals connects to ground and the other outside terminal connects to 5 volts.*

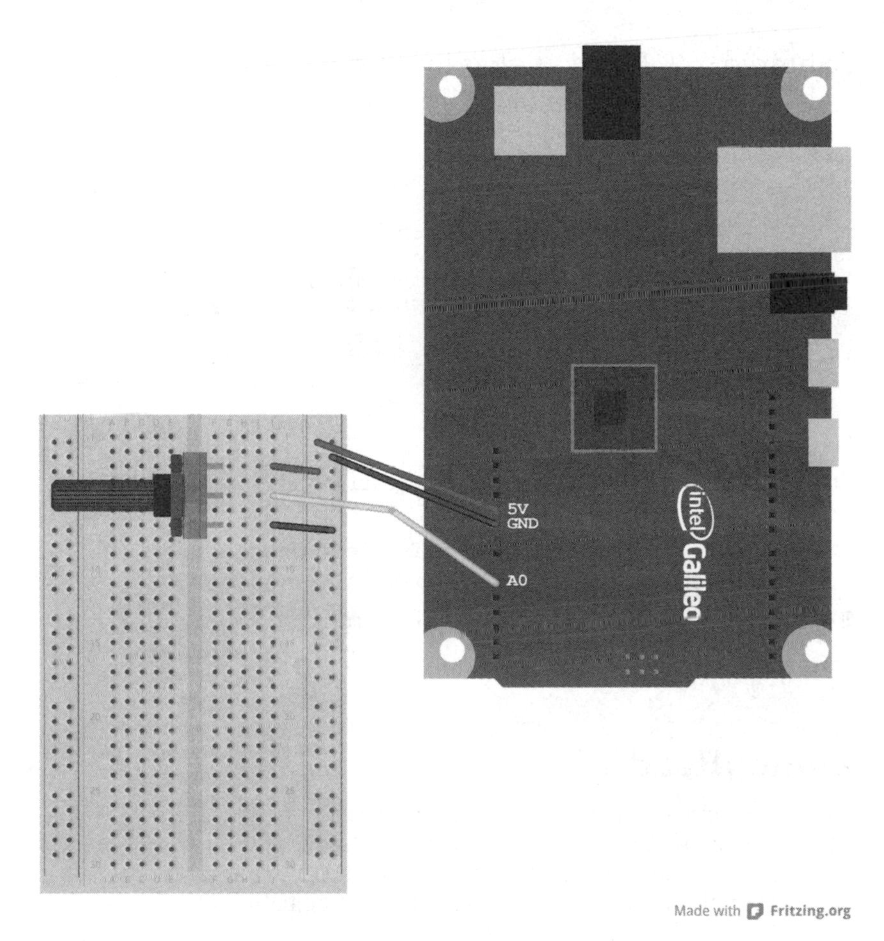

Figure 4-8. *A diagram of how the potentiometer connects to Galileo*

Example 4-2. Basic analog input sketch

```
const int potentiometerPin = 0; // ❶

void setup() {
  Serial.begin(9600);
}

void loop() {
  int sensorReading = analogRead(potentiometerPin); // ❷
  int displayValue = map(sensorReading, 0, 1023, 0, 100); // ❸
  Serial.print("Potentiometer is at ");
  Serial.print(displayValue);
  Serial.println("%.");
  delay(500);
}
```

❶ Create a constant integer called `potentiometerPin` and assign it the value 0.

❷ Create a new variable integer called `sensorReading` and store the analog reading from `potentiometerPin` in it.

❸ Scale the analog input value (which runs from 0 to 1023) to a percentage (0-100%). Store that value in a new variable called `displayValue`.

analogRead()

When you explored "digitalRead()" on page 73, the function returned a value, either high or low. Like `digitalRead()`, `analogRead()` returns a value as well. However, the value is an integer between 0 and 1023. That value represents the amount of voltage going into the analog input pin, from 0 to 5 volts.

That return value can be stored in a variable for use later in your code or it can be evaluated "on the spot," such as within an `if` statement's condition:

```
if (analogRead(0) > 1000) {
        Serial.println("Ludicrous Speed GO!")
}
```

In Example 4-2, the value from `analogRead` is stored in a variable called `sensorReading`.

The syntax of `analogRead()` is:

```
analogRead(pin)
```

The parameter is:

- `pin`: the analog pin number to read (0 to 5).

`analogRead()` returns an integer between 0 and 1023 (representing 0 to 5 volts).

If you changed the IOREF jumper so that you're using 3.3 volts, you must not send more than 3.3 volts to the analog input pins. It's also important to know that the returned value from `analogRead()` will still be on the scale of 0 to 1023 to represent 0 to 5 volts, even if IOREF is set to 3.3 volts.

Code and Syntax Notes

Like previous examples, Example 4-2 also introduces a few more programming concepts to help you along your way.

Constants

To store the pin numbers in memory, you previously used variables. However, in Example 4-2 you tried something a little different:

```
const int potentiometerPin = 0;
```

This syntax creates a spot in memory for an integer called `potentiometerPin` and stores 0 in it, just like a variable. Except this isn't a variable, it's a *constant*. After you assign it the value when you initialize it, you cannot change the value again. If you try to, you'll receive an error when you compile the sketch.

This might be helpful if you want to store a value in memory and you know that it should never be changed. Since the compiler won't compile it and will return an error, you'll know when you've done something wrong in your code.

map()

Because the return value of `analogRead()` is a value between 0 and 1023, you'll sometimes need to scale that value to another range (see Figure 4-9). In the case of Example 4-2, you used the `map()` function to scale that to a percentage (0 to 100) with the following code:

```
int displayValue = map(sensorReading, 0, 1023, 0, 100);
```

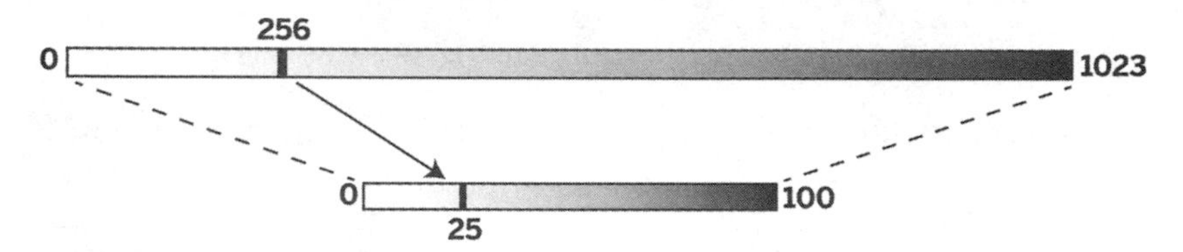

Figure 4-9. *The map function is responsible for scaling a value from one set of ranges to another. In the preceding example, if the input scale went from 0 to 1023 and the output scale went from 0 to 100, inputting the value 256 would return 25.*

Knowing what the result of this line of code is without looking at any reference information, you might be able to tell what each parameter is. If not:

The syntax of `map()` is:

```
map(input, inFrom, inTo, outFrom, outTo)
```

The parameters of `map()` are:

- `input`: the input value to be scaled
- `inFrom`: the first number in the input scale
- `inTo`: the second number in the input scale
- `outFrom`: the first number in the output scale
- `outTo`: the second number in the output scale

`map()` returns a value on the scale of `outFrom` to `outTo`.

To work with `map()` on your own, try modifying Example 4-2 so that it outputs the approximate voltage going into the analog input pin. You can also try taking analog input reading and remap it to an analog output pin.

Variable Resistors

While there are many that do, not all analog sensors work like the potentiometer does, which provides a variable voltage based on some factor (such as the amount the dial on the pot is turned).

Some sensors are simply variable resistors that resist the flow of electricity based on some factor. For instance, a photocell like the one in Figure 4-10 will act like a resistor that changes values based on the amount of light hitting the cell. Add more light and the resistance goes down. Take away light and the resistance goes up. The force sensitive resistor decreases its resistance as you put pressure on the pad. You'll learn more about this in "Force Sensitive Resistor" on page 84.

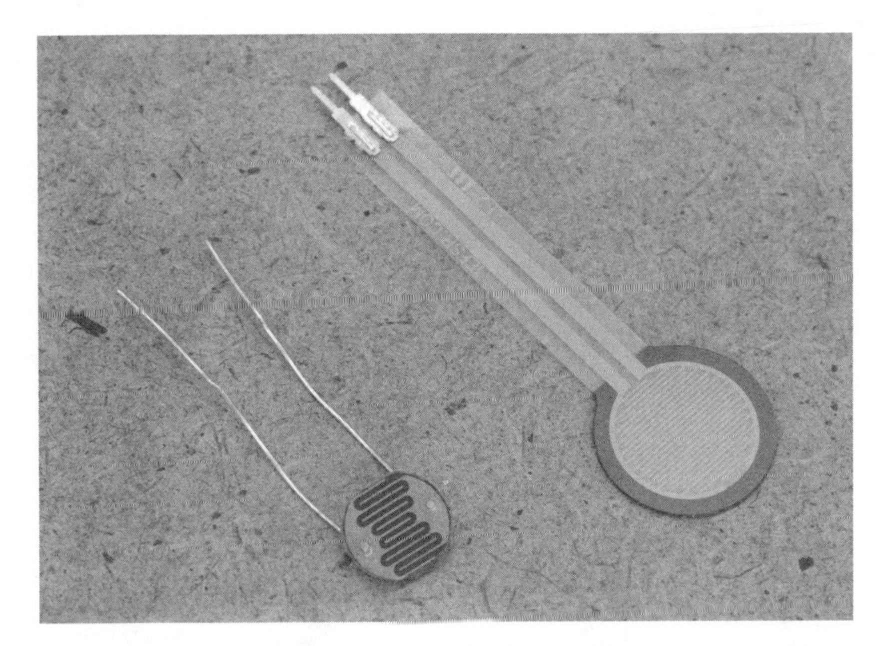

Figure 4-10. *The photocell and force sensitive resistors act as variable resistors and can be used as analog inputs.*

In order to read sensors like these with Galileo's analog input pins, you'll need to incorporate a *voltage divider* circuit.

Voltage Divider Circuit

When you're working with sensors that offer variable resistance, the purpose of a voltage divider is to convert the variable resistance into a variable voltage, which is what the analog input pins are measuring. First take a look at a simple voltage divider.

In Figure 4-11, you'll see two resistors of the same value in series between the positive and ground and one wire to analog input 0 from between the two resistors. Since both resistors are 10K ohms, there's 2.5 volts going to analog input 0.

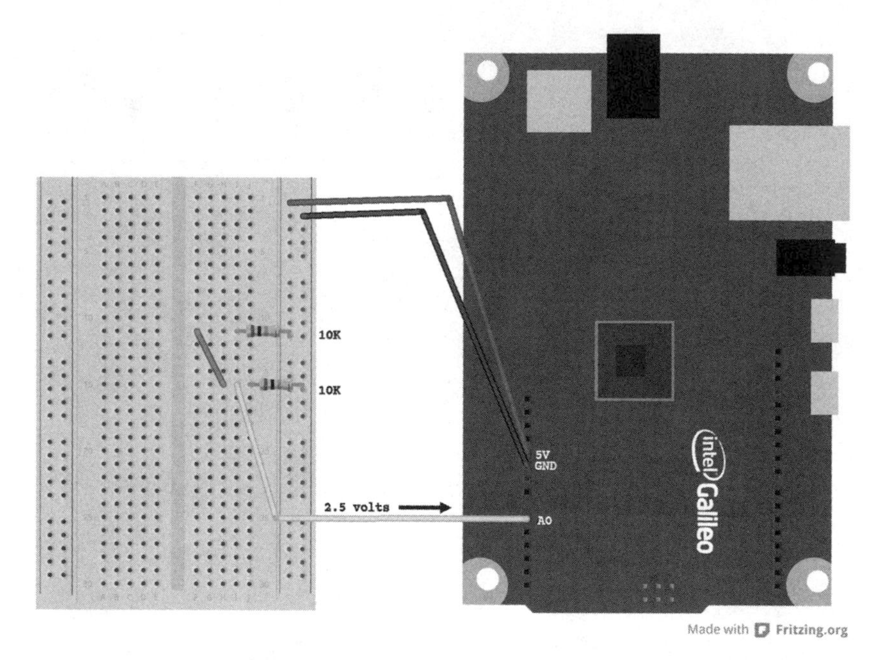

Figure 4-11. *With two of the same value resistors between voltage and ground, the voltage between the two would be half of the total voltage.*

Without getting bogged down in the math involved, if you removed the 10K resistor connected to 5 volts and replaced it with a resistor of a higher value, the voltage going into the analog pin would decrease. If you removed that 10K resistor and replaced it with one of a lower value, the voltage going into the analog pin would increase. We can use this principle with sensors that are variable resistors in order to read them with the analog input pins. You'll simply replace one of the resistors with your sensor.

To try the circuit out, you'll wire up a type of variable resistor called a *force sensitive resistor*, or FSR.

Force Sensitive Resistor

A force sensitive resistor is a variable resistor that changes based on the amount of pressure placed on its pad. When there's no pressure on the pad, the circuit is open. When you start placing pressure on it, the resistance drops.

The exact figures will depend on your particular FSR, but typically you'll see 100K ohms of resistance with light pressure and 1 ohm of resistance with maximum pressure. If you have a *multimeter*, you can measure the changes in resistance to see for yourself, or you can look at the component's *datasheet* which will tell you what to expect from the sensor.

If you're going to replace the resistor connected to 5 volts in Figure 4-11 with a variable resistor like an FSR, you'll want the value of the other resistor to be somewhere in between the minimum and maximum resistance so that you can get the most range out of the sensor. For a typical FSR, try a 10K ohm resistor. Here's what you'll need to try this out:

- Breadboard (Makershed.com part number MKEL3, Adafruit.com part number 64, Sparkfun.com part number 12002)
- Jumper wires (Makershed.com part number MKSEEED3, Adafruit.com part number 758, Sparkfun.com part number 08431)
- Force sensitive resistor (Adafruit.com part number 166, Sparkfun.com part number 09375)
- 10K ohm resistor (Assorted resistor packs: Makershed.com part number MKEE4, Sparkfun.com part number 10969)

To connect the FSR to Galileo:

1. Insert the FSR into a breadboard so that each lead is in a different row.
2. Connect one side of the FSR to Galileo's analog input pin 0.
3. Connect that same side to ground using a 10K resistor. It will have the color bands brown, black, orange, and then gold or silver.
4. Connect the other side of the FSR to 5 volts from Galileo. Your circuit should look like Figure 4-12 or Figure 4-13.
5. Upload the code from Example 4-3 to the board.
6. Open the serial monitor.

Watch the serial monitor as you squeeze the sensor's pad.

Figure 4-12. *How to connect a force sensitive resistor to Galileo*

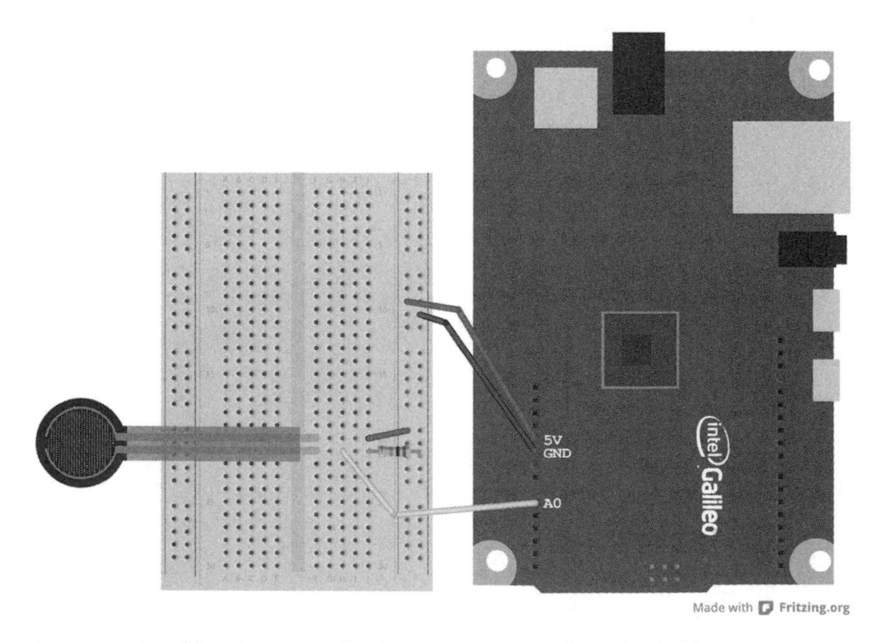

Figure 4-13. *How to connect a force sensitive resistor to Galileo*

Example 4-3. Reading a force sensitive resistor

```
#define FSR_PIN 0 // ❶

void setup() {
  Serial.begin(9600);
}

void loop() {
  int sensorReading = analogRead(FSR_PIN); // ❷
  if (sensorReading < 10) { // ❸
        Serial.println("I don't feel much at all!");
  }
  else if (sensorReading < 600) { // ❹
        Serial.println("Thanks for the squeeze!");
  }
  else { // ❺
        Serial.println("Ouch!");
  }
  delay(1000);
}
```

❶ Tell the compiler to replace all instances of `FSR_PIN` with 0 before compiling the sketch.

❷ Take the analog reading from `FSR_PIN` (pin 0) and store it in an integer variable called `sensorReading`.

❸ If the `sensorReading` is less than 10, print "I don't feel much at all!" via serial.

❹ Otherwise, if `sensorReading` is less than 600, print "Thanks for the squeeze!" via serial.

❺ Otherwise, print "Ouch!" via serial. Based on the previous `if` statements, this would only be executed if `sensorReading` is greater than or equal to 600.

Code and Syntax Notes

Not only did you use `analogRead()` to try out the FSR in Example 4-3, there are also a couple of new coding concepts: `#define` and `else if`.

#define

`#define` is considered a *preprocessor directive*. It tells the compiler to do a "find and replace" before it compiles. In the case of Example 4-3, all instances of `FSR_PIN` are replaced with `0` and then the code is compiled.

Preprocessor directives like `#define` do not get a semicolon terminator at the end of the line.

Writing the name of the `#define` in all capital letters is not required, but it's a convention used by most programmers.

Practice using `#defines` by improving the code in Example 4-3. The threshold between feeling nothing and feeling a squeeze is 10. The threshold between feeling a squeeze and saying "ouch" is 600. Try setting these up as `#define` statements so it's easier to adjust the values according to your sensor.

else if

With `else if`, you can check for another condition if the first `if` condition evaluates as false. The syntax looks like this:

```
if (condition A) {
        execute this code if condition A is true
}
else if (condition B) {
        execute this code if condition A is false
        and condition B is true.
}
else {
        execute this code if both condition A
        and condition B are false.
}
```

You can chain together many `else if`s together and optionally you can have an `else` statement at the end of the chain for code that should be executed if all the previous conditions evaluate as false.

Going Further

Now that you've explored outputs and inputs with Galileo, you can create a device that's interactive. Here are a few ideas for how you can apply what you've learned in this chapter:

- Create a reaction timer game that tests how fast you can press a button after seeing an LED turn on. The scores can be sent via serial to the computer or displayed in an array of LEDs.
- Experiment with reading the temperature with Adafruit's tutorial on how to use the TMP36 Temperature Sensor (*http://bit.ly/1ghLQD6*).
- Create a pattern memory game like Simon (*http://bit.ly/1eouNLu*) using LEDs and buttons.

Here are some additional resources if you'd like to learn more about and explore the concepts covered in this chapter:

- My friend Marc de Vinck made a great video about digital input with Arduino (*http://bit.ly/1i3cqQr*). The same concepts apply to using digital input on Galileo and other hardware development boards.
- There are so many different types of sensors available to you. Check out the selection available at Adafruit (*http://bit.ly/1nFWE06*), Sparkfun (*http://bit.ly/OhCVac*), and Maker Shed (*http://bit.ly/1qsS87m*).
- You can experiment with voltage divider circuits with this online calculator (*http://bit.ly/1lSjQtu*).

5/Going Further with Code

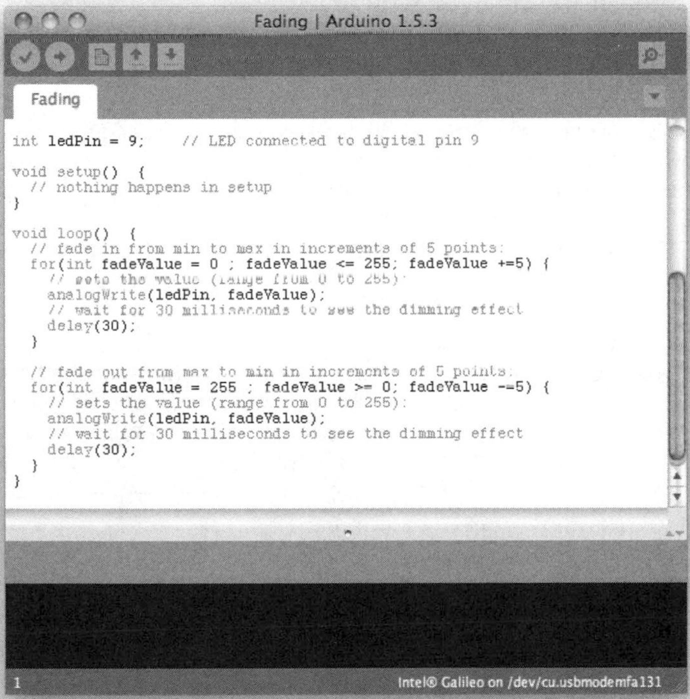

Fading | Arduino 1.5.3

Fading

```
int ledPin = 9;    // LED connected to digital pin 9

void setup()  {
  // nothing happens in setup
}

void loop()  {
  // fade in from min to max in increments of 5 points:
  for(int fadeValue = 0 ; fadeValue <= 255; fadeValue +=5) {
    // sets the value (range from 0 to 255):
    analogWrite(ledPin, fadeValue);
    // wait for 30 milliseconds to see the dimming effect
    delay(30);
  }

  // fade out from max to min in increments of 5 points:
  for(int fadeValue = 255 ; fadeValue >= 0; fadeValue -=5) {
    // sets the value (range from 0 to 255):
    analogWrite(ledPin, fadeValue);
    // wait for 30 milliseconds to see the dimming effect
    delay(30);
  }
}
```

1 Intel® Galileo on /dev/cu.usbmodemfa131

As you learned about how to work with inputs and outputs on Galileo, you also got a tutorial on a few different coding concepts. This short chapter aims to fill in some of the gaps so that you'll have the tools you need to create your own projects.

Data Types

When creating variables, so far the only *data type* you've encountered is the integer, which is meant for storing whole numbers, or in the case of Example 4-1, the digital pin states HIGH or LOW. But there are plenty of other types of data that you can store in memory with Galileo.

int

Since you're already familiar with the basics of the integer, I'll give you a little more information about it. Because an integer on Galileo is stored with 32 binary bits including one bit for the sign (positive or negative), you can count up to the integer 2,147,483,647. If you add 1 to that number, it rolls over to -2,147,483,648.

If 2,147,483,647 isn't high enough for you, you can also use an *unsigned integer*, which takes away the bit needed to determine whether it's positive or negative. An unsigned integer can count from 0 to 4,294,967,295. If you add 1 to that maximum, it will roll over to 0. Initializing a new unsigned integer is as easy as initializing a regular integer:

```
unsigned int bigNumber = 4294967295;
```

Other Arduino boards have different ranges for the type integer. For instance, an integer on Arduino Uno can range between -32,768 and 32,767. Keep this in mind if you're porting code to or from other boards.

Another important thing to know about integers is that if you do a math operation such as division and expect a number with a decimal, the result will simply be the whole number with the decimal chopped off. That's right, if you were to divide two integers and normally expect a result of 2.75, the result you'll actually get is 2! Try 11 divided by 4 on a calculator and in your code:

```
Serial.println(11/4);
// integer division result: 2
```

float

If you do need to work with decimals, the data type `float` is there for you. Its name is short for *floating point number*. As with `int`s, you can store `float` variables in memory. Here's how:

```
float cost = 29.95;
```

And if you want to do math with decimals, you'll simply use a decimal point to indicate that you'd like to do math with floating points so that you don't lose the decimal like you do with integers:

```
Serial.println(11.0/4);
```

The line above would print 2.75 to the serial monitor.

long

When looking at Arduino code from the official examples or from other projects, you may encounter the data type `long`. This type is mostly helpful for Arduinos like the Uno, where integers use fewer bits (16 bits versus 32 on the Galileo) and therefore can't count as high. `long` is a type meant to store a wider range of numbers than `int` and use less precious memory. However, on the Galileo, both `int` and `long` (along with their `unsigned` counterparts) use the same number of bits and therefore have the same range.

While using an `int` will suffice in most cases, it's important to be aware of the existence of `long` as some Arduino functions return this data type. For instance, see "millis()" on page 95. It's also important to be aware of `long` if you are writing a sketch that you may run on an Arduino Uno, Leonardo, or compatible board. On those boards, the maximum value for an `int` is 32,767.

boolean

A `boolean` variable is meant to store the values `true` or `false`. This data type is frequently used to store "flags" to indicate a state or mode of your sketch. You can then evaluate those variables in loops and `if` statements. For instance:

```
#define START_BUTTON 3

boolean gameStarted = false; // ❶

void setup() {
        pinMode(START_BUTTON, INPUT);
}

void loop() {
        if (digitalRead(START_BUTTON)) {
                gameStarted = true; // ❷
        }
        if (gameStarted) { // ❸
                // game play code here
        }
}
```

1. Create a new boolean variable called `gameStarted` and store `false` in it.
2. If the start button is pressed, set `gameStarted` to `true`.
3. If `gameStarted` is true, execute the code in the block.

Using flags like this can help you structure your sketch so that you can check for a few conditions at one time and then act on those conditions later in your code.

char

The data type `char` is a byte that represents an *ASCII* character. ASCII is a system from the early days of computers for translating between bytes and characters. Typically, you'll use the `char` data type when you're reading the characters that are sent to Galileo via serial (see "More Serial" on page 99). Here's how to create and assign a `char` variable:

```
char letter = 'm';
```

When using a character in your code (for instance, when you assign it to a variable or when you compare variable to a particular value), it will always be between single quotation marks:

```
char letter = 'm';

if (letter == 'm') {
        // This code will be executed
}
```

String Object

If you need to work with text and not just single characters, you'll want to familiarize yourself with the `String` object. It's a little different from the previous data types because it's technically not a variable, but an object, much like the Servo object in Chapter 3. Because it's an object, it means that there are special things a `String` can do that other variables cannot.

You may notice that I'm using a capital S when referring to `String` objects. This is a convention to distinguish it from an array of `char`s, which is considered a string (with a lower case *s*). While character arrays use less memory, they're a little bit harder to work with than `String` objects.

Here's one way to make a new `String` object:

```
String gatekeeperString = "There is no Dana, only Zuul!";
```

There are many other ways to create `String` objects and to work with them. Check out the Arduino Reference (*http://arduino.cc/en/Reference/StringObject*) for more information.

millis()

Keeping track of time can be very helpful in an Arduino sketch. In Example 3-1, you tried out the `delay()` function, which stops a sketch for the specified number of milliseconds. But what if you just want to set up a few things to happen on different intervals? That's just one way that `millis()` comes in handy.

The `millis()` function returns the number of milliseconds since the sketch started. You can then check that time against the time you expect something to execute. If the current time is past the time you expected it to execute, then you execute the code and set the next time you expect it to execute.

To explore this concept further, read through the Arduino example BlinkWithoutDelay. From within the IDE, click File → Examples → 02.Digital → BlinkWithoutDelay.

Other Loops

In Chapter 3 you first encountered the `loop` function, which executes its block of code repeatedly until the board is reset or power is disconnected. But what if you want something to happen repeatedly *within* the main `loop` function? Or what if you want to loop through a bunch of pins within the `setup` function? There are a few different types of loops available to you for just this purpose.

while

A `while` loop simply executes a block of code over and over again as long as the condition is true. The syntax is:

```
while (condition) {
        execute this code repeatedly as long as condition is true
}
```

If the condition isn't true when Galileo gets to that point in the code, it simply won't execute the block of code at all.

do... while

A `do... while` loop, on the other hand, executes a block of code and *then* checks for a condition. If that condition is true, it executes the block of code again. This continues until the condition evaluates as false.

```
do {
        execute this code once and then continue executing it
        repeatedly as long as the condition is true
} while (condition)
```

This means that the code inside a `do... while` block will always execute at least once.

for Loops

A `for` loop is a way of having a block of code execute a specific number of times. The syntax is a bit complicated at first, but you'll find yourself using them and encountering them a lot more than `while` and `do... while` loops. With practice, you'll be writing and reading `for` loops fluently.

Let's start by taking a look at a basic `for` loop:

```
for (int i = 0; i < 10; i++) {
    Serial.println("Hello, Galileo!");
}
```

The code above would print "Hello, Galileo!" ten times to the serial monitor.

The `for` loop statement has three parts inside the parentheses and each part is separated by a semicolon. First the *initialization*, which is code that is executed once before anything else happens. The second part is the *condition*. It's evaluated and if the statement is true, then the code in the block is executed. After the code in the block is executed, the third part of code in the `for` loop statement, the *afterthought*, is executed. To summarize the syntax:

```
for (initialization; condition; afterthought) {
    execute this code if condition is true
}
```

To explore what's going on, take a look at another `for` loop:

```
for (int i = 0; i < 3; i++) {
    Serial.print("Loop iteration number: ");
    Serial.println(i);
}
```

This would print:

```
Loop iteration number: 0
Loop iteration number: 1
Loop iteration number: 2
```

Here's a breakdown of what's going on, to explain how that `for` loop works:

- Create a new variable called `i` and set its value to zero. This initialization is defined by the first part of the `for` statement.
- Check if `i` is less than 3. Because it's 0, this evaluates to be true. This condition is defined by the second part of the `for` statement.
- Print "Loop iteration number: 0" to the serial monitor.
- Increment `i` by one, so that its value is now 1. This afterthought is defined by the third part of the `for` statement.
- Check if `i` is less than 3. Because it's 1, this evaluates to be true.
- Print "Loop iteration number: 1" to the serial monitor.
- Increment `i` by one, so that its value is now 2.
- Check if `i` is less than 3. Because it's 2, this evaluates to be true.
- Print "Loop iteration number: 2" to the serial monitor.
- Increment `i` by one, so that its value is now 3.
- Check if `i` is less than 3. Because it's 3, this evaluates to be false and the loop ends.

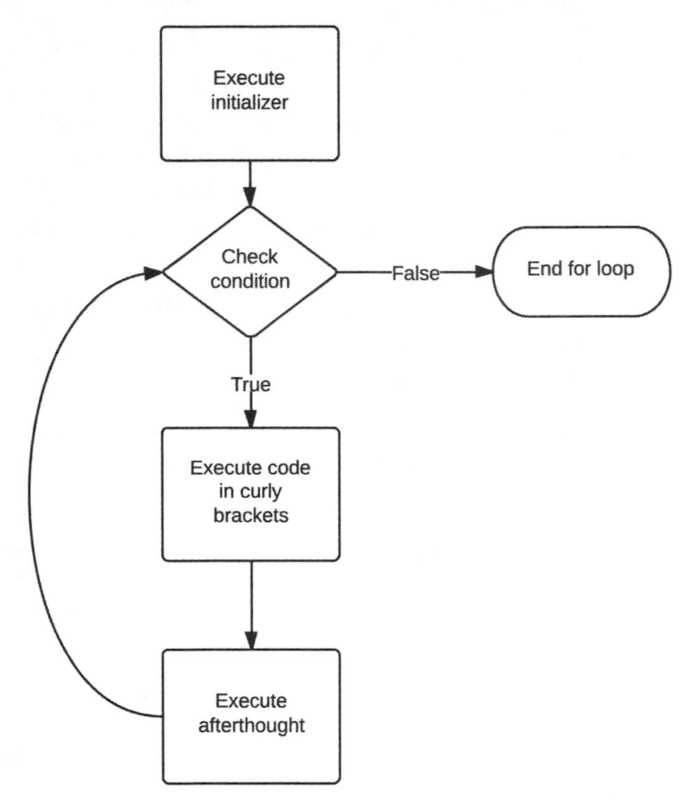

Figure 5-1. *This flow chart is another way of showing how a `for` loop works.*

In the `for` loop above, you'll notice that the loop iterated 3 times, but it only counted up to the number 2. As a typical convention in programming, you start counting from the number 0. Therefore, a list with 10 items would be numbered 0 through 9.

The most common way that a `for` loop is used is to have a particular block of code run a specific number of times. You'll therefore see a `for` loop like this quite frequently:

```
for (int i = 0; i < 10; i++) {
    // This will be executed 10 times
}
```

As with "Local Variables" on page 74, the variable `i` in the example above can only be accessed within the looping block of code. After the `for` loop has executed its last iteration, the variable will be destroyed.

More Serial

In Chapter 2, you learned how Galileo can send data via serial to a computer. But serial can act as a two-way channel of communication between your Galileo and computer (or other device). There are a few `Serial` functions that help you process the data that's sent to your board.

Serial.available() and Serial.read()

On the Galileo, there's a *serial buffer* which stores all the bytes of serial data it receives. When you use your sketch to read a byte from serial, you're actually reading the first byte that's "at the front of the line" in the buffer. After reading that byte, it's removed from the buffer, and the next byte (if there is one) is ready to be read next.

In other words, the serial buffer follows a *first in, first out*, or FIFO, convention. The first byte received into the buffer is the first byte out when read by your sketch.

There is a limited amount of space for data in the serial buffer, so you want to make sure that your sketch is reading the data at least as frequently as the transmitting device is sending it.

If the device that's transmitting bytes sends a lot of data and your sketch doesn't call `Serial.read()` frequently enough, the buffer will *overflow* and you may get unexpected results.

One way to avoid this problem is to use a system of "call and response." When the Galileo is ready for data, it sends a byte to the other device. The other device will respond with some data. The Galileo can then take its time processing the data and acting on it. When it's ready for more data, it will send another byte.

To try out reading serial information, upload Example 5-1 to your board and open the serial monitor.

Example 5-1. Serial receive example

```
#define LED 13

void setup() {
```

```
        pinMode(LED, OUTPUT);
        Serial.begin(9600);
}

void loop() {
        if (Serial.available()) { // ❶
                char c = Serial.read(); // ❷
                if (c == 'h') {
                        digitalWrite(LED, HIGH);
                }
                if (c == 'l') {
                        digitalWrite(LED, LOW);
                }
        }
}
```

❶ If there are characters available in the serial buffer, execute the code in the block.

❷ Create a new `char` variable called `c` and store the next byte from the serial buffer in it. This will remove it from the serial buffer.

In the Arduino IDE, open the serial monitor by clicking the magnifying glass button on the upper right side of the window. In the input text box at the top of the serial monitor, enter a lowercase h or l and click Send. The on-board LED connected to pin 13 should turn on when you send the "h", and it should turn off when you send an "l." Sending any other character should have no effect on the LED.

The function `Serial.available()` returns the number of bytes in the serial buffer. If there are none, the `if` statement in Example 5-1 will evaluate as false and the code in the block won't be executed. Any value other than 0 will evaluate as true.

The function `Serial.read()` will return the next byte in the serial buffer, which removes it from the buffer.

Taking It Further

With everything this book has covered so far, you should be able to understand most of the examples that are included with the Arduino IDE. However, it can take some practice to really cement these concepts. Here are a few ideas to help you do just that:

- To learn more about loops and other control structures, take a look at the examples included with the Arduino IDE that explore these concepts. Click File→Examples→05.Control.
- One of the control examples relates to the `switch` and `case` keywords, which work together to help you evaluate for a few different conditions

at once, much like chaining a bunch of `if` statements together. To try it out, click File→Examples→05.Control→switchCase in the IDE.

- To learn more about serial communication, take a look at the examples under File→Examples→04.Communication in the IDE. Many of the examples demonstrate how to make Arduino interact with your computer and include code for Processing (*http://processing.org/*), which is a simple programming language and development environment. Much like Arduino, it makes getting started really easy. Not only that, but the IDE should feel familiar to you because the Arduino IDE is based on the Processing IDE!
- Try writing a text adventure game (*http://en.wikipedia.org/wiki/Interactive_fiction*) that you can play via serial. Maybe it can also incorporate some physical components. Perhaps you could light a real lamp when the player instructs your game to `LIGHT LAMP`.

6/Getting Online

Internet connectivity is an important feature of today's hardware development boards. Galileo not only has built-in network connectivity, but it also does a fantastic job supporting sketches that use the standard Arduino Ethernet and WiFi libraries (this means that Galileo can run the same sketches you run on other Arduino boards). And if you choose to tap into the power of Linux, you can do even more with an Internet connection outside of your sketch.

In this chapter, you will use Galileo to:

- Connect to a remote server to get information

- Have your Arduino code use some of the Linux capabilities of Galileo
- Act as a web server to serve information to a web browser

Connecting and Testing an Ethernet Connection

First you should make sure that you can get your board to connect to the Internet in the simplest way possible. Try it out now:

1. Connect your Galileo via an Ethernet cable to your network. You can either plug it directly into an available port on your router, or if you are wired for Ethernet, plug it into an active jack.

Some large networks—like those at companies, schools, and hotels—don't let any device just plug in and connect to the Internet. For instance, when I wanted to start using the Galileo at New York University, I had to use my computer's web browser to log in to my school account and register Galileo's MAC address as a device that I own. If you need to know your Galileo's MAC address, it can be found either on the back side of the board (on the Mini PCI Express port) or on top of the Ethernet port on the top of the board.

2. Connect your Galileo to power.
3. Connect your computer to Galileo via the USB client port.
4. Within the Arduino IDE, select File→Examples→Ethernet→WebClient.
5. Click Upload.
6. Open the serial monitor.

If your board was successful in connecting to a server, you'll see text start to appear in the serial monitor (Figure 6-1). This example programs your Galileo to do a Google search for the term "Arduino." As the HTML response from Google's server is received, it sends those characters to the Serial Monitor.

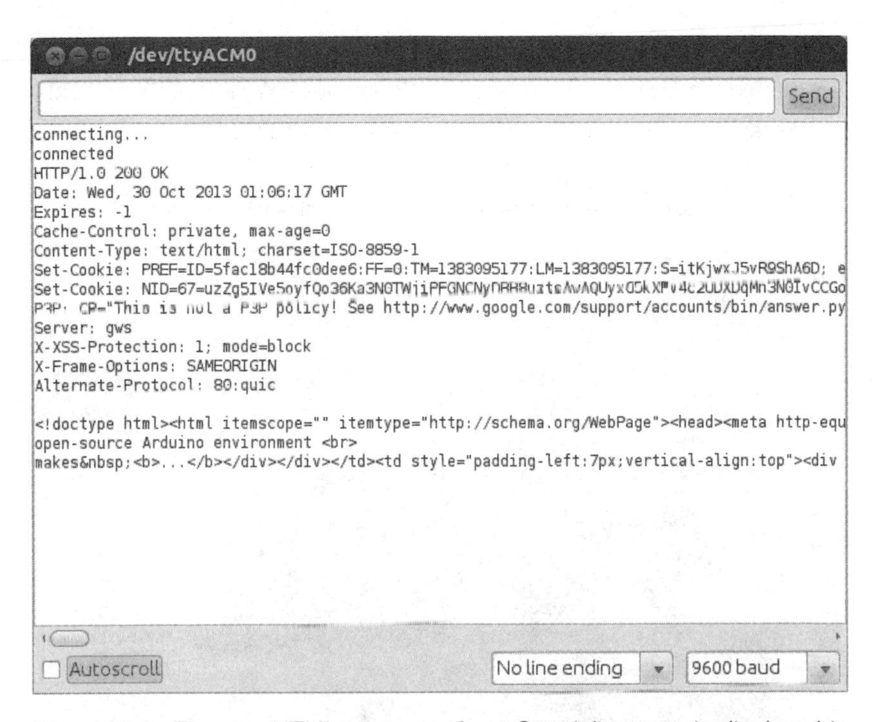

Figure 6-1. *The raw HTML response from Google's server is displayed in the serial monitor. A lot of the text may appear on a single line. While the text is coming in, you can turn off Autoscroll in the serial monitor and scroll back to the beginning.*

Connecting and Testing with a WiFi Connection

On the back side of the Galileo is a Mini PCI Express slot that can hold a WiFi module so that your project can connect to the Internet wirelessly.

In order to use a WiFi card, you must boot Galileo off of a MicroSD card that has a more complete version of Linux than runs on Galileo by default. This is because the drivers needed to work with WiFi Mini PCI Express cards do not fit into Galileo's on-board flash memory. For information on how to load a MicroSD card with Linux and have Galileo boot off of it, see Appendix D.

To test your wireless connection:

1. With the board disconnected from USB and power, insert your WiFi module on the bottom of the board (Figure 6-2).
2. Be sure to attach an antenna if your WiFi module requires it.
3. Connect your Galileo to power.

4. Connect your computer to Galileo via the USB client port.
5. Within the Arduino IDE, select File→Examples→WiFi→WiFiWebClient.
6. Enter your WiFi network's SSID and password on the lines:

```
char ssid[] = "yourNetwork";       // your network SSID (name)
char pass[] = "secretPassword";    // your network password
```

7. Click Upload.
8. Open the serial monitor.

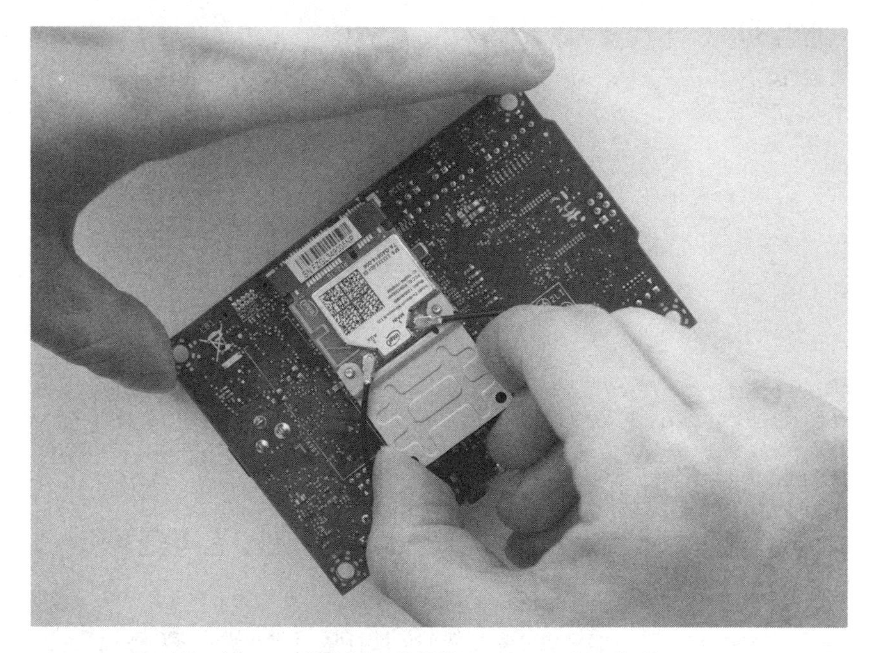

Figure 6-2. *Attaching a WiFi Mini PCI Express card to Galileo*

If your board was successful in connecting to a server via WiFi, you'll see text start to appear in the serial monitor. The first few lines will be information about your WiFi connection, such as the name of the SSID, the IP address your router assigned to your Galileo, and the signal strength. After that, you'll see a raw HTML response from Google's servers. As noted in the caption of Figure 6-1, you may want to turn Autoscroll off so that you can look at the text as it's coming in.

Connecting Using Linux Commands

Now that you've tested your connection using the Ethernet library or WiFi library for Galileo, I want to introduce you to a different way to make connections. As mentioned in Chapter 1, there's a version of Linux running on Galileo, and a simple bit of Arduino code can run any Linux command. You tried this out a bit in "Looking at Linux" on page 60, but here you're going to take a deeper dive.

If you've ever used the Linux or Unix command line before, you'll feel pretty comfortable with this. Those of you who are well-versed in the command line will quickly see how powerful this can be. No matter what your experience level is, you'll find this to be a handy trick when you want to do things that are more complicated than Arduino code can handle easily.

1. Make sure your Galileo is connected to the network via Ethernet, or has the WiFi module installed and tested.
2. Connect power to the Galileo.
3. Connect your computer to Galileo via the USB client port.
4. Create a new sketch within the Arduino IDE and enter in the code from Example 6-1.
5. Upload the code to the Galileo.
6. Open the serial monitor.
7. In the text entry field at the top of the serial monitor, type any character and click Send.

If everything is working correctly, you should see HTML from MAKE's website pouring into your serial monitor.

Example 6-1. Using a system call to connect to a server

```
void setup() {
  Serial.begin(9600);
}

void loop() {
  if (Serial.available()) { // ❶
    Serial.read(); // ❷
    system("curl http://makezine.com &> /dev/ttyGS0"); // ❸
  }
}
```

❶ If there's a character available in the serial buffer, execute the code in the block below. Where does this character come from? You'll type it into the serial monitor.

❷ Read a character from the serial buffer in order to remove it.

❸ Execute the Linux command `curl` with the URL `makezine.com` and push the output back to the serial monitor.

The Serial functionality in Example 6-1 is included so that you can tell Galileo to make a single request, as opposed to just executing it once when the sketch starts (easy to miss in the serial monitor) or executing it repeatedly (which wouldn't be very kind to the server, and might get your network temporarily blocked because it looks like a network attack).

system()

The `system()` function is a special one for Galileo. It tells your Arduino code to execute a command-line operation in the Linux shell. To learn how it works, take a closer look at its use in Example 6-1:

```
system("curl http://makezine.com &> /dev/ttyGS0");
```

The `system()` function itself takes a single parameter, a string. It's the exact command that you would type on the command line. In this case, we're using the command `curl` to fetch the URL *http://makezine.com/*. So that we can see if the command works, the output of that request should be redirected back to our serial monitor. The `&>` indicates that standard output (normal text output) and errors (most error messages) should be redirected to `/dev/ttyGS0`, which is the device that Galileo uses to display information in the Arduino IDE's serial monitor.

Getting Galileo's IP Address Using system()

From time to time, you may need to know information about how your Galileo is connected to the network. Frequently, I want to know a device's IP address so that if the board is acting like a server, I can connect to it.

The command-line function `ifconfig` allows you to see information about your connections. You can try it out on your computer's command line if you're running Linux or OS X. On Windows, the similarly named `ipconfig` fills this role.

To execute `ifconfig` on the Galileo through Arduino code and see the results in the serial monitor:

1. Make sure your Galileo is connected to the network via Ethernet, or has the WiFi module installed and tested.
2. Connect power to the Galileo.

3. Connect your computer to Galileo via the USB client port.
4. Create a new sketch within the Arduino IDE and enter in the code from Example 6-2.
5. Upload the code to the Galileo.
6. Open the serial monitor.

Example 6-2. Using a system call to see output of ifconfig

```
void setup() {
  Serial.begin(9600);
}

void loop() {
  system("ifconfig &> /dev/ttyGS0"); // ❶
  delay(3000);
}
```

❶ Execute the command `ifconfig` and output the results to the serial monitor.

If you see text like the text below in the serial monitor, you'll know you have it working correctly (if you're connected with WiFi, the `wlan0` adapter will appear in the list, too).

```
eth0      Link encap:Ethernet  HWaddr 00:13:20:FD:F6:5D
          inet addr:10.0.1.119  Bcast:0.0.0.0  Mask:255.255.255.0
          inet6 addr: fe80::213:20ff:fefd:f65d/64 Scope:Link
          UP BROADCAST RUNNING MULTICAST  MTU:1500  Metric:1
          RX packets:2625 errors:0 dropped:0 overruns:0 frame:0
          TX packets:877 errors:0 dropped:0 overruns:0 carrier:0
          collisions:0 txqueuelen:1000
          RX bytes:632810 (617.9 KiB)  TX bytes:143590 (140.2 KiB)
          Interrupt:41 Base address:0x4000

lo        Link encap:Local Loopback
          inet addr:127.0.0.1  Mask:255.0.0.0
          inet6 addr: ::1/128 Scope:Host
          UP LOOPBACK RUNNING  MTU:65536  Metric:1
          RX packets:0 errors:0 dropped:0 overruns:0 frame:0
          TX packets:0 errors:0 dropped:0 overruns:0 carrier:0
          collisions:0 txqueuelen:0
          RX bytes:0 (0.0 B)  TX bytes:0 (0.0 B)
```

The preceding example lists two network devices: the first, `eth0`, represents the Ethernet connection to the router. The second, `lo`, is a *local loopback*. It's a virtual network device for testing and enables programs running on your device to connect to local services (or *daemons*) even if there's no Ethernet or WiFi connection.

The device we want to focus on is `eth0` (or `wlan0` if you are using WiFi). Within the information for that device, the IP address is indicated after `inet addr:`. In the case of my Galileo, it's 10.0.1.119. If I want to connect *to* my Galileo from another device on my network, I'll use that address. I could even have my router expose that IP to the Internet if I want to access it from wherever I have an Internet connection (after taking some security precautions of course). See "Connecting to Galileo from the Internet" on page 124 for more information on that.

Connecting to Servers

When you browse the Web with your web browser, you're seeing a human-readable display of information, hopefully laid out in a way to make it easy to consume the content you want. When machines need to get content from the web, however, they don't necessarily need all the layout and design information—just the data.

Many sites that provide services such as weather, social networking, communication, and file storage also offer an *application programming interface*, or API. It's the way that information is communicated to or from a particular service.

For instance, if you wanted your site or device to post a photo to your Facebook profile, you'd use Facebook's Graph API (*https://developers.facebook.com/docs/graph-api/*). You can also use the Graph API to download a list of your friends. These APIs are Facebook's way of saying, "here's how to have your devices and services communicate with our servers."

However, these services can be complex to use (for example, just authenticating yourself to the service can take a lot of programming), so we're going to stick to a very simple example of an API to start off.

How Many Days Until MAKE Comes Out?

I've created a simple website called How Many Days Until MAKE Comes Out? (*http://nextmakemagazine.appspot.com/*), which serves one simple purpose: it tells you how many days until the next issue of MAKE Magazine is expected to hit newsstands. The data is entered into a database by me based on the production calendars I receive as a MAKE contributing editor. It's served by a free tier of Google App Engine, and the source code for the site is available on GitHub (*http://bit.ly/1kJKPXR*) if you want to take a look at how I made it.

When you go to *http://nextmakemagazine.appspot.com/* in your web browser, you'll see that the information is formatted to be viewed and understood by a human, but the server is also configured to speak directly to simple microcontrollers by stripping away all the extra style and language and only returning the number of hours until the next issue is released. You can see this if you go to *http://nextmakemagazine.appspot.com/simple*.

Your Galileo can use its Internet connection via Ethernet or WiFi to connect to this URL, take the data it receives, and evaluate how to display it (see Figure 6-3). First, let's make sure it can connect to the server.

1. Walk through the instructions in "Connecting and Testing an Ethernet Connection" on page 104 or "Connecting and Testing with a WiFi Connection" on page 105 to make sure you have connectivity to the Internet.
2. Create a new sketch and enter the code from Example 6-3.
3. Upload the code and then open the serial monitor.

Figure 6-3. *Galileo has many ways to connect physical objects to the Internet!*

In the serial monitor, you should see a number printed every five seconds. This number represents the number of hours until the next issue of MAKE is released (in case you want to be more precise!).

Example 6-3. Getting simple data from the Internet

```
void setup() {
  Serial.begin(9600);
}
```

```
void loop() {
  Serial.println(getHours()); // ❶
  delay(5000);
}

int getHours() { // ❷
  char output[5]; // ❸
  FILE *fp; ❹
  fp = popen("curl http://nextmakemagazine.appspot.com/simple", "r"); // ❺
  if (fp == NULL) { // ❻
    Serial.println("Couldn't run the curl command.");
    return -1;
  }
  else { ❼
    fgets(output, sizeof(output), fp);
  }
  if (pclose(fp) != 0) {  // ❽
    Serial.println("The curl command returned an error.");
    return -1;
  }
  return atoi(output); // ❾
}
```

❶ Call the function `getHours()` and print its result. `getHours()` is defined below.

❷ Create a new function called `getHours` that will return an integer.

❸ Create an array of characters called `output` for storing the response.

❹ Create a *file pointer* called `fp`, which is how our code will reference the output of the Linux command.

❺ Use the Linux `curl` command to fetch the number of hours and store it in `fp`.

❻ If there was a problem running `curl`, report the error in the serial monitor and have the `getHours()` function return -1.

❼ Otherwise, read the data in `fp` and put it into the `output` array.

❽ If `curl` had a problem getting the data (for instance, if the server is not available), report the error in the serial monitor and have the `getHours()` function return -1.

❾ Have the function return the contents of the array as an integer using the built-in `atoi` (ASCII character array/string to integer) function.

Defining Functions

The first thing you might notice in Example 6-3 is that the `loop` function has only two lines of code. The `delay(5000)` ensures that each iteration of the loop happens only every five seconds. But what about

`Serial.println(getHours());`? The innermost function, `getHours()`, is actually defined right below the loop function.

The first line of the function definition indicates that our function is called `getHours` and will return an integer as a result. Whenever the function is called from the `setup` or `loop` functions, the code inside the function is executed. For `getHours`, it will request data from the server, store that response, and then convert that response into an integer value representing the number of hours you'll need to wait until a new issue of MAKE is available. That integer is returned from the function as long as there are no errors. If there are any errors, it will return the value -1.

Executing a Linux command and getting its response is just one way that you can get data into your sketch. This is a very unique Galileo feature because it's a powerful way to connect different technologies together. You can also have data passed between the Linux system and your Arduino code by writing and reading files. See "Another Approach to Passing Data" on page 117 for a few hints on how to do that.

Converting ASCII Characters to Integers

There's another interesting thing going on in Example 6-3. Let's say the server sent a response of 45. It's actually sending two ASCII characters, a 4 and a 5. When the Arduino reads these characters, it's not going to understand them as the integer 45, but rather as two bytes that represent the characters 4 and 5. This means if you need to do math with the value, you must convert them to an integer. That's what the function `atoi()` is for. It will look at the array of characters and output their value as an integer, which you can use for arithmetic.

Parsing JSON with Python

The How Many Days Until MAKE Comes Out? (*http://nextmakemagazine.appspot.com/*) server makes the preceding example pretty easy because it supplies one simple piece of data. Other services may provide a few different pieces of data structured in a format called *JSON*, or JavaScript Object Notation.

JSON has become the standard format for transmitting structured data through the web. If you want to be able to read data from a site that offers JSON, you'll have to parse it. As this would be difficult to do with Arduino code, you can use other languages to do this job and pass the appropriate information into Arduino's code.

Another format for structuring data is called *XML*, or *eXtensible Markup Language*.

To preview JSON data, visit *http://nextmakemagazine.appspot.com/json* in your web browser.

The response will likely be together on one line, but if you were to add line breaks and indentation, it would look like Example 6-4. There are three key/value pairs: the number of hours until the next issue, the next volume number, and the number of days until the next issue.

Example 6-4. JSON response

```
{
  totalHours: 1473,
  volumeNumber: "38",
  daysAway: 61
}
```

The code in Example 6-5 uses the Python programming language to connect to the server's JSON feed at *http://nextmakemagazine.appspot.com/json* and parses the volume number and number of hours.

Example 6-5. Using Python to parse JSON

```
import json # ❶
import urllib2 # ❷

httpResponse = urllib2.urlopen('http://nextmakemagazine.appspot.com/json') # ❸
jsonString = httpResponse.read() # ❹

jsonData = json.loads(jsonString) # ❺

print "Volume", jsonData['volumeNumber'], "will be released in", \
      jsonData['totalHours'], "hours." # ❻
```

❶ Import Python's JSON library in order to parse the JSON response.

❷ Import Python's urllib2 library to fetch the data from the server.

❸ Connect to the server to request the JSON feed and store the response in `httpResponse`.

❹ Store the body of the response in `jsonString`.

❺ Convert the string into a Python data object.

❻ Print the information out.

There are a few different ways to put Python code onto the board. In this section, you'll connect to the board and enter the code into the text editor, vi.

In order to use Python and save your files on the board, you must boot Galileo with an SD card. See Appendix D for more information on how to create a bootable MicroSD card.

1. Connect to Galileo's command line using Telnet (see "Connecting via Telnet" on page 61) or serial (see Appendix H).
2. Change to root's home directory:

```
# cd /home/root/
```

3. Launch the text editor vi with the filename *json-parse.py* to create that file.

```
# vi json-parse.py
```

4. Along the left side of the screen, you'll see a column of tildes (~). Type the letter i to enter insert mode. An "I" will appear in the lower lefthand corner of your screen.
5. Enter the code from Example 6-5 into vi.
6. Hit the escape key to switch from insert mode to command mode. The "I" in the lower left corner will disappear and you'll see a dash instead.
7. Type *:x* and press Enter to save the file and exit vi.
8. Test the script by executing the code from the command line:

```
# python json-parse.py
```

If you got everything right, you should see the following output on the command line:

```
Volume 38 will be released in 1473.0 hours.
```

Unlike with Arduino code, Python is a bit particular about how each line of code is indented. As long as you don't add any leading spaces or tabs before the lines of code in Example 6-5, you'll be fine. In Example 6-11 later in this chapter, you'll see an example where you'll have to be careful with your leading white space.

As you can see from Example 6-5, parsing the JSON response from a website isn't very difficult when you have Python available to you on Galileo. Now you simply need to connect the response from Python to your Arduino code.

To try that now, first modify *json-parse.py*:

1. On Galileo's command line, be sure you're still in root's home directory:

```
# cd /home/root/
```

2. Open the file for editing in vi:

```
# vi json-parse.py
```

3. Type the letter i to enter insert mode. An "I" will appear in the lower left hand corner of your screen.
4. Edit the file so that it reflects the code in Example 6-6.
5. In the Arduino IDE, create a new sketch with the code from Example 6-7. You'll see that it's very similar to Example 6-3. Instead of calling `curl` from the command line, it uses Python to run the script you wrote from Example 6-6.
6. Upload the code to the board.

Example 6-6. Using Python to parse JSON

```
import json
import urllib2

httpResponse = urllib2.urlopen('http://nextmakemagazine.appspot.com/json')
jsonString = httpResponse.read()

jsonData = json.loads(jsonString)

print jsonData['daysAway'] # ❶
```

❶ Print only the number of days until the next MAKE will come out from the JSON response.

Example 6-7. Calling Python from Arduino code

```
void setup() {
  Serial.begin(9600);
}

void loop() {
  Serial.println(getDays());
  delay(5000);
}

int getDays() {
  char output[5];
  FILE *fp;
  fp = popen("python /home/root/json-parse.py", "r");  ❶
  if (fp == NULL) {
    Serial.println("Couldn't run the curl command.");
```

```
    return -1;
  }
  else {
    fgets(output, sizeof(output), fp);
  }
  if (pclose(fp) != 0) {
    Serial.println("The curl command returned an error.");
    return -1;
  }
  return atoi(output);
}
```

❶ Use Python to execute your script, writing the output to `fp`.

Now, open the Serial Monitor in Arduino. You should see the response from the server as the number of days until the next issue of MAKE comes out.

Another Approach to Passing Data

It's important to know that every time you use the `system()` or `popen()` functions to run Python code, it can take time for Galileo to launch the Python interpreter and run the code. This also means that nothing else in your Arduino code will execute until that process is completely finished and Python has exited.

In this project, it'll work fine. But as your projects grow in complexity, you may not want to wait for Python to relaunch each time you need to fetch data.

To solve this, you can also have your Python code run constantly in the background, updating a file (or files) that the Arduino code will read. In order to get the Arduino code to run your Python script and then move on to the rest of the Arduino code, simply append an ampersand to the end of the system call so that the Python script runs in the background and your Arduino code continues to execute. For instance:

```
system("python /home/root/json-loop.py &");
```

For more information about having Python read and write files, I recommend exercises 15 through 17 of the free online resource, Learn Python the Hard Way (*http://learnpythonthehardway.org/book/*).

As an example of reading files within Arduino code, Example 6-8 shows a modified version of the `getHours()` function in Example 6-3. It has been rewritten so that the response from the server is piped into a file called *response.txt* and then read by the Arduino code.

Example 6-8. Writing and reading a file with Arduino code

```
int getHours() {
  char output[5];
    system("curl  http://nextmakemagazine.appspot.com/simple  >  re
sponse.txt");
  FILE *fp;
  fp = fopen("response.txt", "r");
  fgets(output, 5, fp);
  fclose(fp);
  return atoi(output);
}
```

Connecting an LCD Character Display

What good is this information if it can only be seen in the serial monitor? Let's hook up an LCD display to print out this information.

In addition to the parts you've been using, you'll also need:

- A standard 16x2 LCD character display (such as Makershed.com part number MKAD30, Adafruit.com part number 181, Sparkfun.com part number 00255)
- Breadboard (Makershed.com part number MKEL3, Adafruit.com part number 64, Sparkfun.com part number 12002)
- Jumper wires (Makershed.com part number MKSEEED3, Adafruit.com part number 758, Sparkfun.com part number 08431)
- 10k ohm potentiometer (Adafruit.com part number 562, Sparkfun.com part number 09288)

To connect the LCD to Galileo:

1. Disconnect your Galileo board from your computer's USB port and from power.
2. Insert the LCD into the breadboard (solder header pins onto it if necessary).
3. Insert the potentiometer into the breadboard as well.
4. Using jumper wires, connect the potentiometer and LCD to Galileo's power and ground as shown in Figure 6-5.
5. Using jumper wires, connect the LCD data lines to Galileo's digital pins as shown in Figure 6-6. Your project might look something like Figure 6-4.
6. Connect power to Galileo.

7. Connect Galileo to your computer via USB.
8. Upload the code from Example 6-9.

Figure 6-4. *Connecting the LCD character display to Galileo*

Example 6-9. Displaying results on an LCD

```
#include <LiquidCrystal.h> // ❶

LiquidCrystal lcd(12, 11, 5, 4, 3, 2); // ❷

void setup() {
  lcd.init(1,12,255,11,5,4,3,2,0,0,0,0); // ❸
  lcd.begin(16, 2); // ❹
  lcd.setCursor(3, 0); // ❺
  lcd.print("days until"); // ❻
  lcd.setCursor(0, 1); // ❼
  lcd.print("MAKE is here!");
}

void loop() {
  lcd.setCursor(0, 0); // ❽
  lcd.print("   "); // ❾
  lcd.setCursor(0, 0); // ❿
  lcd.print(getDays()); // ⓫
  delay(30*60*1000); // ⓬
}
```

```
int getDays() {
  char output[5];
  FILE *fp;
  fp = popen("python /home/root/json-parse.py", "r");
  if (fp == NULL) {
    Serial.println("Couldn't run the curl command.");
    return -1;
  }
  else {
    fgets(output, sizeof(output), fp);
  }
  if (pclose(fp) != 0) {
    Serial.println("The curl command returned an error.");
    return -1;
  }
  return atoi(output);
}
```

1. Include the code from Arduino's Liquid Crystal library.
2. Create a new object called `lcd` and specify which Galileo pins are connected to it.
3. Initialize the LCD using the pin numbers from before. Thanks to Github user houmei for this hint: *https://gist.github.com/houmei/8505883*. This is only needed with the Galileo and isn't needed for a regular Arduino.
4. Specify that the LCD has two lines of 16 characters each.
5. Set the cursor to the fourth character on the first line. (Start counting from 0.)
6. Print the message "days until".
7. Reposition the cursor to the first character on the second line. (Again, counting from 0.)
8. Reset the cursor to the beginning of the first line.
9. Clear out anything that's there by printing three spaces.
10. Reset the cursor to the beginning of the first line again.
11. Print the response from our Python script, which parses the JSON response from the server.
12. Wait 30 minutes before refreshing the display.

After uploading the code, the message should appear in the LCD display! You might need to adjust the display's contrast using the potentiometer. It should look something like Figure 6-3.

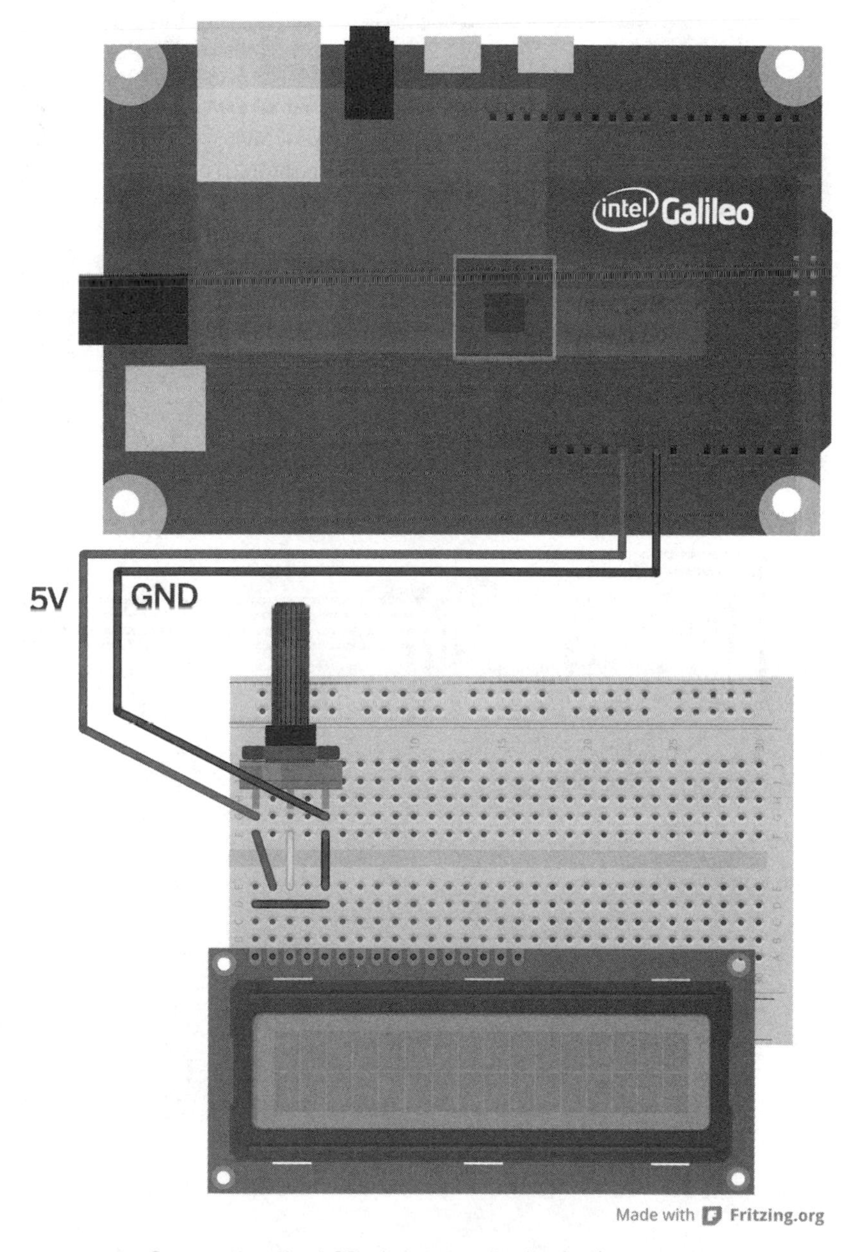

Figure 6-5. *Connecting the LCD character display to the potentiometer and the Galileo's 5-volt and ground pins*

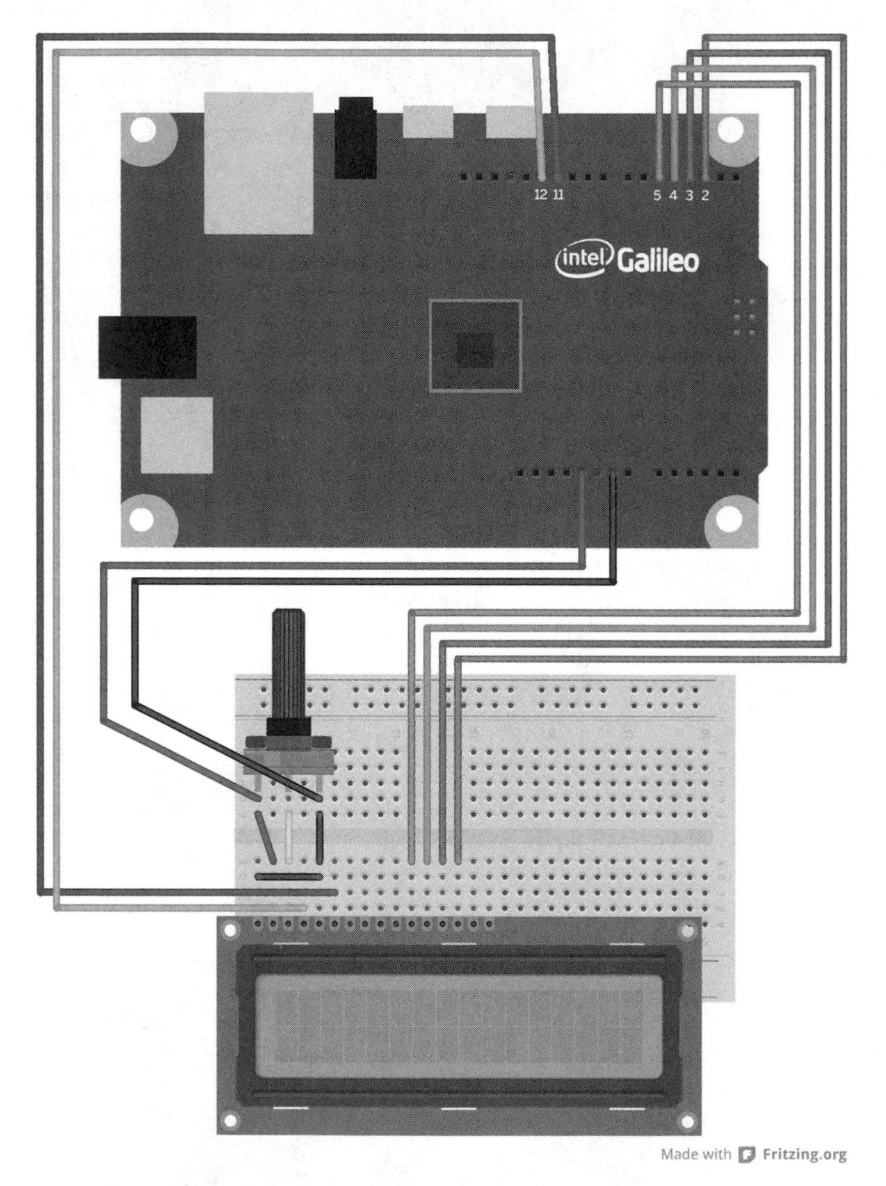

Figure 6-6. *Connecting the LCD's data pins to the Galileo's digital I/O pins*

Serving a Web Page

Just as the Galileo can contact a web server to get or post information, it can also act as a simple web server itself. You can use your web browser to connect to it and get information about its input pins (such as sensors you've connected) or have it actuate an output when a browser connects. To test out these capabilities, you're going to have a web page update when a button is pressed.

1. Just as you did in Figure 4-2, wire up a button to pin 2 of the Galileo.
2. Connect the Galileo to your network via Ethernet.
3. Create a new sketch with the code from Example 6-10 and upload it to your board. This example is based on the Arduino example found in File → Examples → Ethernet → WebServer.
4. Open the serial monitor to see the IP address that was assigned to your Galileo by your router.
5. Open a web browser, enter that IP into the address bar, and hit Enter.

You should see the following text on screen: "The button is not pressed!" Try pushing the button and hitting refresh on your browser. You should then see "The button is pressed!" (Figure 6-7)

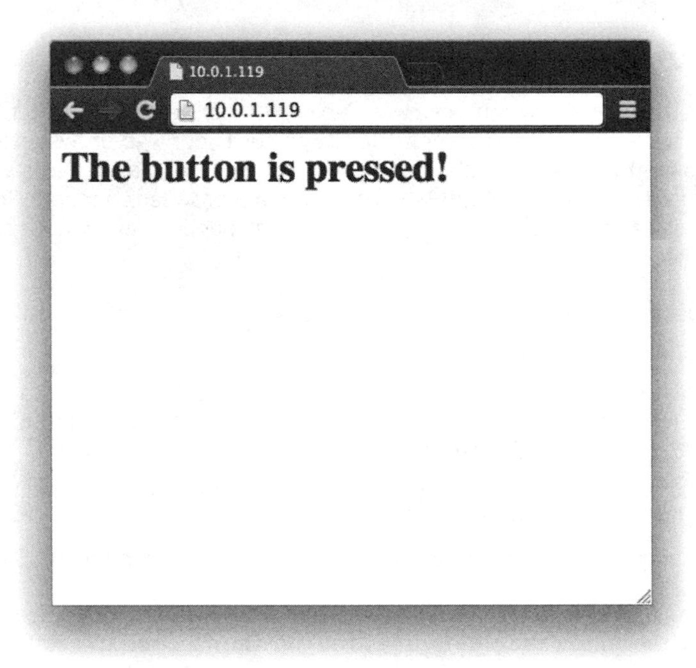

Figure 6-7. *The response from Galileo displayed in a web browser*

Connecting to Galileo from the Internet

Depending on your network setup, this IP address is probably only accessible by computers that are on the same network. If you'd like to make Galileo accessible over the Internet, you'll need to adjust your router's settings so that it assigns your Galileo the same IP every time it powers on (look for something called *MAC address reservations* in your router's settings) and you'll also need to have connections from the Internet passed through your router's firewall to the Galileo (look for a setting called *port forwarding*).

Example 6-10 shows how Arduino code can respond to requests from your web browser to show information from a button, but it could also show sensor information from around your house. Using Arduino code to create a server can be an acceptable solution for some simple things.

Example 6-10. A basic web server

```
#include <Ethernet.h>
int buttonPin = 2;
byte mac[] = { 0xDE, 0xAD, 0xBE, 0xEF, 0xFE, 0xED };
```

```
EthernetServer server(80); // ❶

void setup() {
  Serial.begin(9600);
  Ethernet.begin(mac);
  server.begin();
  Serial.print("server is at ");
  Serial.println(Ethernet.localIP());
  pinMode(buttonPin, INPUT);
}

void loop() {
  EthernetClient client = server.available();
  if (client) { // ❷
    Serial.println("new client");
    boolean currentLineIsBlank = true; // ❸
    while (client.connected()) {
      if (client.available()) { // ❹
        char c = client.read(); // ❺
        Serial.write(c); // ❻
        if (c == '\n' && currentLineIsBlank) { // ❼
          client.println("HTTP/1.1 200 OK"); // ❽
          client.println("Content-Type: text/html");
          client.println("Connection: close");
          client.println();
          client.println("<!DOCTYPE HTML>"); // ❾
          client.println("<html>");
          if (digitalRead(buttonPin)) { // ❿
            client.println("<h1>The button is pressed!</h1>");
          }
          else { // ⓫
            client.println("<h1>The button is not pressed!</h1>");
          }
          client.println("</html>");
          break;
        }
        if (c == '\n') {
          currentLineIsBlank = true;
        }
        else if (c != '\r') {
          currentLineIsBlank = false;
        }
      }
    }
    delay(1);
    client.stop();
    Serial.println("client disconnected");
  }
}
```

❶ Have the server listen on port 80, the default port for HTTP (web) communication.

❷ If a client connects, execute the code in the following block of code.

❸ Create a flag to indicate when there's a blank line in the request from the client.

❹ If data from the client's request is available, execute the code that follows.

❺ Read in the request from the client.

❻ Echo the request to the serial monitor.

❼ When we receive two blank lines in a row, the client is ready to receive data.

❽ Start sending the HTTP headers.

❾ Start sending HTML.

❿ If the button is pressed, send a piece of HTML that indicates that this is the case.

⓫ If the button is not pressed, send the alternate piece of HTML.

Serving a Web Page with Python

Even without Arduino code, you can have Python read from and write to Galileo's pins. You can therefore use Python alone to do the same thing that Example 6-10 does. These instructions show you how.

1. Just as you did in Figure 4-2, wire up a button to pin 2 of the Galileo.
2. Connect to Galileo's command line using Telnet (see "Connecting via Telnet" on page 61) or serial (see Appendix H).
3. Change to root's home directory:

```
# cd /home/root/
```

4. Launch the text editor vi with the filename *server.py* to create that file.

```
# vi server.py
```

5. Along the left side of the screen, you'll see a column of tildes (~). Type the letter `i` to enter insert mode.
6. Enter the code from Example 6-11 into vi.
7. Hit the escape key to switch from insert mode to command mode. The "I" in the lower-left corner will disappear and you'll see a dash instead.
8. Type `:x` and press Enter to save the file and exit vi.
9. From the command line, execute the script:

```
# python server.py
```

10. Open a web browser, enter your Galileo's IP into the address bar and hit Enter.

You should see the text on screen: "The button is not pressed!" Try pushing the button and hitting refresh on your browser. You should then see "The button is pressed!"

Example 6-11. Serving web pages with Python

```
import SocketServer
import SimpleHTTPServer # ❶

PORT = 80 # ❷
class MyTCPServer(SocketServer.TCPServer):
  allow_reuse_address = True # ❸
class myHandler(SimpleHTTPServer.SimpleHTTPRequestHandler):
  def do_GET(self):
    self.send_response(200) # ❹
    self.send_header("Content-type", "text/html")
    self.end_headers()
    self.wfile.write("<html><body>") # ❺
    self.wfile.write("<h1>The button is")
    with open("/sys/class/gpio/gpio32/value", "r") as gpio: # ❻
      state = gpio.read(1) # ❼
      if state == "0": # ❽
        self.wfile.write(" not") # ❾
      self.wfile.write(" pressed!</h1>") # ❿
      self.wfile.write("</body></html>")
httpd = MyTCPServer(("", PORT), myHandler)
print "Serving from port", PORT
httpd.serve_forever()  # ⓫
```

❶ Import the Python libraries for server functionality.

❷ Set the port to 80, the default for HTTP.

❸ Avoid "address already in use" error.

❹ Send the HTTP response headers.

❺ Send the HTTP response body.

❻ Open the GPIO system file for pin 2 to get the pin's status.

❼ Store one byte from that file into `state`.

❽ If the state is 0, it means the button is not pressed (low).

❾ Reply with the word "not" if the button is not pressed.

❿ Send the rest of the HTTP response.

⓫ Run the server until the user presses Control-C.

Be sure to get the indentation just the way it's shown in Example 6-11. You can use the tab button or spaces as long as the number of spaces for each level of indentation is consistent. That is to say, indenting by one level is two spaces, indenting by two levels is four spaces, and so on.

If you look at your terminal connection Galileo, you'll probably see something like this for each time you refresh your browser:

```
192.168.2.1 - - [01/Jan/2001 04:39:50] "GET / HTTP/1.1" 200 -
192.168.2.1 - - [01/Jan/2001 04:39:51] "GET /favicon.ico HTTP/1.1" 200 -
```

The first line is showing your browser's request for the HTML page. The second request also comes from your browser. It's looking for a graphic file called `favicon.ico`. It's a file that many web sites serve so that its icon will appear in the address bar or in your bookmarks.

Type Control-C to terminate the server.

Starting on Boot

You can make Galileo execute that Python script every time it boots.

1. From the command line, change to the directory */etc/init.d*:

   ```
   # cd /etc/init.d
   ```

2. Use vi to create a new file called *start-server.sh*. This will be a *shell script*, which is just a series of commands saved in a file.

   ```
   # vi start-server.sh
   ```

3. Type `i` to enter insert mode and add the following line, which will execute the server script with Python and supress any output:

   ```
   python /home/root/server.py >> /dev/null 2>&1 &
   ```

4. Press escape, then `:x`, then Enter to save the file and exit vi.
5. From the command line, make the script executable:

   ```
   # chmod +x start-server.sh
   ```

6. Then add that script as a system service:

   ```
   # update-rc.d start-server.sh defaults
   ```

7. Now shut down the board:

```
# shutdown -h now
```

8. Disconnect the USB cable and power, and then reconnect the power.

Now every time the Galileo launches, your server will start.

If you want to disable the script from starting up again, simply execute:

```
# update-rc.d -f start-server.sh remove
```

Taking It Further

In this chapter, you used Galileo's Linux capabilities to connect to Internet servers and even created your own basic server with Arduino code. Here are a few ways that you can push your Galileo even further:

- If you want to learn more about what Python is capable of, the free online course Learn Python the Hard Way (*http://learnpythonthehardway.org/*) is a fantastic resource.
- When you boot Galileo off the SD card, it also includes Node.js (*http://nodejs.org/*), a platform for creating server applications with the language JavaScript. This can be another powerful tool for creating dynamic web apps that run off of Galileo. Just like with Python, Node.js can work with the pins directly, or communicate with your Arduino code.
- A few folks from Intel created ConnectAnyThing (*https://github.com/IntelOpenDesign/ConnectAnyThing*), "an open platform intended for easy prototyping of connected appliances, devices, and installations using Intel's Galileo development board." It's cool because it turns Galileo in a WiFi hotspot and lets you read inputs and control outputs from any browser.

A/Arduino Code Reference

analogRead

Gets the voltage value of a particular analog input pin. The values range from 0 to 1023, which represents the voltages 0 to 5. See "Analog Input" on page 75.

If the IOREF jumper is set to 3.3V, then the maximum input voltage for the analog input pins is 3.3 volts. However, this will not adjust the scale of `analogRead()`, which will always be based on 0 to 5 volts.

Syntax

```
analogRead(pin);
```

Parameters

`pin`
: The pin number

Return value

integer (between 0 and 1023, representing 0 to 5 volts)

Example

```
int sensorReading = analogRead(0);
```

Gets the value from analog pin 0 and stores it in a new integer variable called `sensorReading`.

analogReadResolution

Sets the resolution of the value returned by `analogRead()`. Galileo's analog-to-digital converter has a resolution of up to 12 bits, but Arduino sets the default value to 10 bits.

By default, `analogRead()` will return values from 0 to 1023. If you call `analogReadResolution(12)`, then `analogRead()` will return values of 0 to 4095. Both ranges represent the same voltage ranges on the analog input pins.

Syntax

```
analogReadResolution(bits);
```

Parameters

`bits`
: The resolution that `analogRead()` will return in bits. The default is 10 and the maximum on Galileo is 12.

Return Value

None

Example

```
analogReadResolution(12);
```

Sets the analog-to-digital converter on Galileo to its maximum resolution, so that `analogRead()` will return a value between 0 and 4095 to represent 0 to 5 volts.

analogWrite

This sets the duty cycle of a pin capable of pulse width modulation. In other words, it will pulse the pin really quickly and allow you to adjust the amount of time it's turned on versus turned off. On the Galileo, only pins 3, 5, 6, 9, 10, and 11 can be used with `analogWrite()`. These pins are marked with a tilde (~) on the board. See "analogWrite()" on page 43.

Syntax

```
analogWrite(pin, value);
```

Parameters

`pin`
: The pin number

`value`
: An integer between 0 (totally off) and 255 (totally on)

Return value

None

Example

```
analogWrite(9, 127);
```

Pulses pin 9 so that it's turned off 50% of the time.

atoi

Converts a string of numbers to an integer. With `atoi`, a string made up of the characters "23" can be converted to the integer 23 to be used in arithmetic. See "Converting ASCII Characters to Integers" on page 113 for more information.

Syntax

```
atoi(stringNumber);
```

Parameters

`stringNumber`
: An array of characters, such as those read from a file.

Return value

The integer value of the string.

Example

```
char number[] = "23";
int i = atoi(number);
```

`number[]` holds an array of characters: 2 and 3. `atoi` converts that into the integer value 23.

const

Store a spot in memory to hold data that cannot be changed. See "Constants" on page 81. This might be helpful if you want to store a value in memory and you know that it should never be changed. Because the compiler will stop and return an error if you try to change a `const` value, you'll know when you've done something wrong in your code.

Example

```
const int potentiometerPin = 0;
```

Stores the integer 0 in a spot in memory called `potentiometerPin`. You will not be able to change this value later in code.

delay

Pauses the execution of code for the specified number of milliseconds. See "delay()" on page 39.

Syntax

```
delay(ms);
```

Parameters

`ms`
: The number of milliseconds to wait

Return value None.

Example

```
delay(1000);
```

This causes the execution of code to stop for one second (1,000 miliseconds).

digitalRead

Reads the value of an input pin and returns high or low. See "digitalRead()" on page 73.

Syntax

```
digitalRead(pin);
```

Return value

HIGH or LOW.

Parameters

`pin`
: The pin number

Example

```
if ( digitalRead(9) == HIGH ) {
    Serial.println("Pin 9 is high.")
}
```

This prints "Pin 9 is high." if it's connected to 5 volts (or 3.3 volts if the IOREF jumper is set to 3.3V).

digitalWrite

Sets the value of a digital pin to high or low (on or off). See "digitalWrite()" on page 38.

Syntax

```
digitalWrite(pin, value);
```

Parameters

`pin`
: The pin number

`value`
: Either HIGH or LOW (case-sensitive)

Return Value

None.

Example

```
digitalWrite(9, HIGH);
```

This turns on pin 9 (sets it to high).

else

Used with the `if` statement, the `else` statement indicates a block of code that should be executed when `if` evaluates as false. See "if... else Statements" on page 74.

Syntax

```
if (condition) {
    //execute this code if condition is true
}
else {
    //execute this code if condition is false
}
```

Example

```
int switchInputPin = 2;

void setup() {
    pinMode(switchInputPin, INPUT);
    Serial.begin(9600);
}

void loop() {
    int switchState = digitalRead(switchInputPin);
    if (switchState == HIGH) {
        Serial.println("The switch is on!");
    }
    else {
        Serial.println("The switch is off!");
    }
    delay (500);
}
```

fclose

This function from the C library is available within Galileo to close an opened file in the Linux file system. See "Another Approach to Passing Data" on page 117 for more information.

Syntax

```
fclose(fp);
```

Parameters

fp
: A pointer to a file.

Return value

If successfully closed, 1. Otherwise, an error occurred.

Example

```
int getHours() {
  char output[5];
  system("curl http://nextmakemagazine.appspot.com/simple > response.txt");
  FILE *fp;
  fp = fopen("response.txt", "r");
  fgets(output, 5, fp);
  fclose(fp);
  return atoi(output);
}
```

fgets

This function from the C library is available within Galileo to read the contents of an opened file in the Linux file system. See "Another Approach to Passing Data" on page 117 for more information.

Syntax

```
fgets(output, bytes, fp);
```

Parameters

output
: An array to store the read bytes

bytes
: The number of bytes to read

fp
: A pointer to a file

Return value

Returns the output bytes read from the file as an array.

Example

```
int getHours() {
  char output[5];
  system("curl http://nextmakemagazine.appspot.com/simple > response.txt");
  FILE *fp;
  fp = fopen("response.txt", "r");
  fgets(output, 5, fp);
  fclose(fp);
  return atoi(output);
}
```

fopen

This function from the C library is available within Galileo to open a file from the Linux file system. See "Another Approach to Passing Data" on page 117 for more information.

Syntax

```
fopen(filename, mode);
```

Parameters

`filename`
: The name of the file to open

`mode`
: "r" to read an existing file, "w" to write a new file or overwrite existing file, "a" to append to a file or create it if it doesn't exist.

Return value

A pointer to a file.

Example

```
int getHours() {
  char output[5];
  system("curl http://nextmakemagazine.appspot.com/simple > response.txt");
  FILE *fp;
  fp = fopen("response.txt", "r");
  fgets(output, 5, fp);
  fclose(fp);
  return atoi(output);
}
```

if

Executes a block of code if a particular condition is true. See "if Statements" on page 48.

Syntax

```
if (condition) {
    //execute this code if condition is true
}
```

Example

```
int n = 10;

if (n > 10) {
    // this will not be executed since n is not greater than 10
    digitalWrite(redLed, HIGH);
}

if (n < 10) {
    // this will not be executed since n is not less than 10
    digitalWrite(greenLed, HIGH);
}

if (n == 10) {
    // this will be executed since n equals 10
    digitalWrite(yellowLed, HIGH);
}
```

int

The data type integer. This creates a spot in memory to hold a single whole number.

Example

```
int led = 13; // Creates a spot in memory called led and stores
              // the number 13 in it.

void setup() {
    pinMode(led, OUTPUT); // the value of led (13) is used to
                          // set the mode.
}
void loop() {
    digitalWrite(led, HIGH);
    delay(1000);
    digitalWrite(led, LOW);
    delay(1000);
}
```

loop

A required function in every Arduino sketch. This is the block of code that's executed repeatedly after the `setup` function. See "Setup and Loop" on page 26.

Example

```
void setup() {
    // The code here will be executed once
    // when the board is booted.
}
void loop() {
    // The code here will be executed
    // repeatedly after setup() is executed.
}
```

map

Scales a value from one range of numbers to another. See "map()" on page 81.

Syntax

```
map(input, inFrom, inTo, outFrom, outTo)
```

Parameters

`input`
: The input value to be scaled

`inFrom`
: The first number in the input scale

`inTo`
: The second number in the input scale

`outFrom`
: The first number in the output scale

`outTo`
: The second number in the output scale

Return value

`map()` returns a value on the scale of `outFrom` to `outTo`.

Example

```
sensorReading = analogRead(0);
int displayValue = map(sensorReading, 0, 1023, 0, 100);
```

Given an input value from analog pin 0, which is on the scale of 0 to 1023, store a new value on the scale of 0 to 100. So if `sensorReading` were 256, `displayValue` would be 25.

pclose

This function from the C library is available within Galileo to close a stream opened by popen. See Example 6-3 for more information.

Syntax

```
pclose(fp);
```

Parameters

fp
: File pointer from a popen stream.

Return value

0 if no error.

Example

```
int getHours() {
  char output[5];
  FILE *fp; ❶
  fp = popen("curl http://nextmakemagazine.appspot.com/simple", "r");
  if (fp == NULL) {
    Serial.println("Couldn't run the curl command.");
    return -1;
  }
  else {
    fgets(output, sizeof(output), fp);
  }
  if (pclose(fp) != 0) {
    Serial.println("The curl command returned an error.");
    return -1;
  }
  return atoi(output);
}
```

pinMode

Sets the direction (or mode) of a digital pin to input or output. See "pinMode()" on page 37.

Syntax

```
pinMode(pin, mode);
```

Parameters

pin
: The pin number

mode
: Either INPUT or OUTPUT (case-sensitive)

Return Value

None.

Example

```
pinMode(13, OUTPUT);
```

This sets pin 13 as an output, which can be controlled with digitalWrite.

popen

This function from the C library is available within Galileo to execute a command in the Linux shell and get its output. See Example 6-3 for more information.

Syntax

```
popen(command, mode);
```

Parameters

command
: The Linux command to execute.

mode
: Usually "r" for reading the response from the command, but can also use "w".

Return value

A pointer to a file.

Example

```
int getHours() {
  char output[5];
  FILE *fp;
  fp = popen("curl http://nextmakemagazine.appspot.com/simple", "r");
  if (fp == NULL) {
    Serial.println("Couldn't run the curl command.");
    return -1;
  }
  else {
    fgets(output, sizeof(output), fp);
  }
  if (pclose(fp) != 0) {
    Serial.println("The curl command returned an error.");
    return -1;
  }
  return atoi(output);
}
```

Serial.begin

Opens the serial port on the Galileo and tells it what speed to send the data. See "Serial.begin()" on page 53.

Syntax

```
Serial.begin(speed);
```

Parameters

speed
: The speed in bits per second (also known as baud). The following standard Arduino speeds can be used: 300, 600, 1200, 2400, 4800, 9600, 14400, 19200, 28800, 38400, 57600, or 115200. With Arduino sketches, 9600 is most common. Intel Galileo also supports speeds of 50, 75, 110, 134, 150, 200, 1800, 230400, 460800, and 500000.

Return value

None.

Example

```
Serial.begin(9600);
```

Opens the serial port and sets the speed to 9600 bits per second.

Serial.print

Transmits data over the serial port. See "Serial.print()" on page 54.

Syntax

```
Serial.print(value);
```

Parameters

value
: The data to send. This can be a string of text, a character, a byte, an integer, or other types of data.

Return Value

long
: The number of bytes transmitted.

Example

```
Serial.print("Hello, world!");
```

Serial.println

Transmits data over the serial port and then sends a carriage return. See "Serial.println()" on page 54.

Syntax

```
Serial.println(value);
```

Parameters

`value`
: The data to send. This can be a string of text, a character, a byte, an integer, or other types of data.

Return Value

long
: The number of bytes transmitted

Example

```
Serial.println("Hello, world!");
```

servo.attach

Assigns a `servo` object to a particular pin. See "Controlling Servos" on page 55.

Syntax

```
myServo.attach(pin);
```

Parameters

`pin`
: The pin number

`myServo`
: Represents any `servo` object

There are additional parameters available for this function. See the full servo library documentation (*http://arduino.cc/en/Reference/ServoWrite*) for details.

Return value

None.

Example

```
Servo myServo;              // create a servo object

void setup() {
    myServo.attach(9);
}
```

servo.write

Sets the position of a servo motor. See "Controlling Servos" on page 55.

Syntax

```
mySer vo.write(angle);
```

Parameters

`angle`
: The angle to set on the servo, from 0 to 180

`myServo`
: Represents any `servo` object

Return value

None.

Example

```
Servo myServo;              // create a servo object

void setup() {
    myServo.attach(9);
    myServo.write(90);
}
```

setup

A required function in every Arduino sketch. This is the block of code that's executed once when your Galileo is booted. See "Setup and Loop" on page 26.

Example

```
void setup() {
    // The code here will be executed once
    // when the board is booted.
}
void loop() {
    // The code here will be executed
    // repeatedly after setup() is executed.
}
```

system

Execute a Linux command. See "system()" on page 108.

`system()` does not work on other Arduino boards.

Example

```
system("curl http://makezine.com &> /dev/ttyGS0");
```

Uses the Linux command `curl` to fetch a site and output the server's response to the Galileo serial monitor.

B/Breadboard Basics

A solderless breadboard helps you make electrical connections between components, but it can be a bit confusing at first. This section will give you a whirlwind tour of the breadboard so that you can understand how to use it while recreating the examples in this book.

In the Figure B-1, the shaded areas show which pins have electrical connections inside the breadboard.

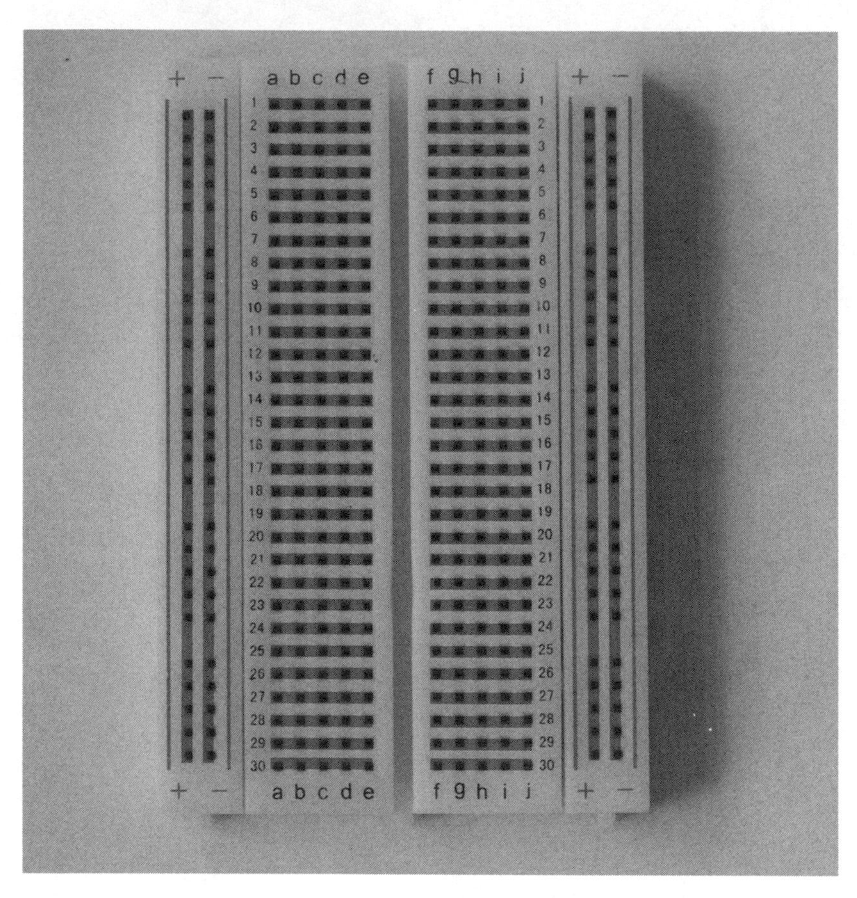

Figure B-1. *Electrical connections are already made between the shaded holes on this solderless breadboard. You'll use jumper wires to connect them to each other and to your Galileo.*

The rails that run up and down the left and right sides are meant for your power and ground connections. If you need power and ground on each side, you'll have to connect each column to power and ground. You can also connect one set of rails to power and ground and then connect each set to each other, as in Figure B-2.

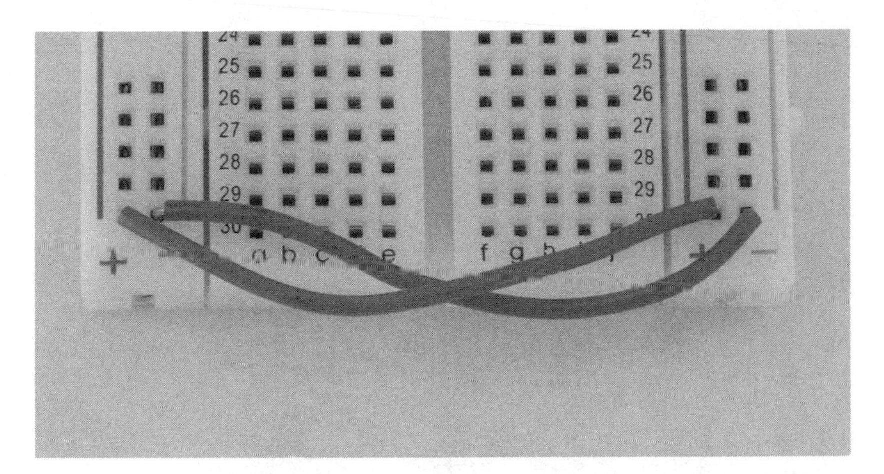

Figure B-2. *Connecting the power rails together sometimes makes things a bit more convenient.*

One thing that frequently trips people up: if you have a full-sized breadboard like the one in Figure B-3, you'll need to connect the top half of the power rails to the bottom half if you want the power to go all the way down to the bottom.

Figure B-3. *On a full-size breadboard like this one, you'll need to connect the top half to the bottom half of the power rails in order to let power and ground run all the way down.*

The breadboard jumpers in Figure B-4 can be very helpful to keep things neat when you start making lots of connections between components.

Figure B-4. *These handy precut jumpers are great for breadboarding.*

C/Resistor Reference

If you're doing basic hobby electronics, you'll typically use resistors with four color bands. The first three bands indicate resistance in ohms, according to Table C-1. The fourth band indicates tolerance, which will usually be gold (5%) or silver (10%), but could also be brown (1%) or red (2%).

The *tolerance* of the resistor indicates the possible amount of variance between the indicated value and the actual value of the resistor. If no tolerance is indicated, it's assumed that it's 20%.

Table C-1. *Basic four-band resistor reference*

Color	First Band	Second Band	Third Band (Multiplier)	Fourth Band (Tolerance)
Black	0	0	x1	
Brown	1	1	x10	1%
Red	2	2	x100	2%
Orange	3	3	x1K	
Yellow	4	4	x10K	
Green	5	5	x100K	
Blue	6	6	x1M	
Purple	7	7	x10M	
Gray	8	8		
White	9	9		
Gold				5%
Silver				10%

To determine the value of a typical resistor, follow these steps:

1. Orient the resistor so that the gold or silver band is on the right side.
2. For the first band, get the first digit of the resistance value from Table C-1.
3. For the second band, get the second digit of the resistance value from Table C-1.
4. For the third band, get the multiplier from Table C-1.

For example, a resistor with the colors brown (1), black (0), orange (1K), and gold (5%), as pictured in Figure C-1, would have the value 10K (10,000) ohms with 5% tolerance.

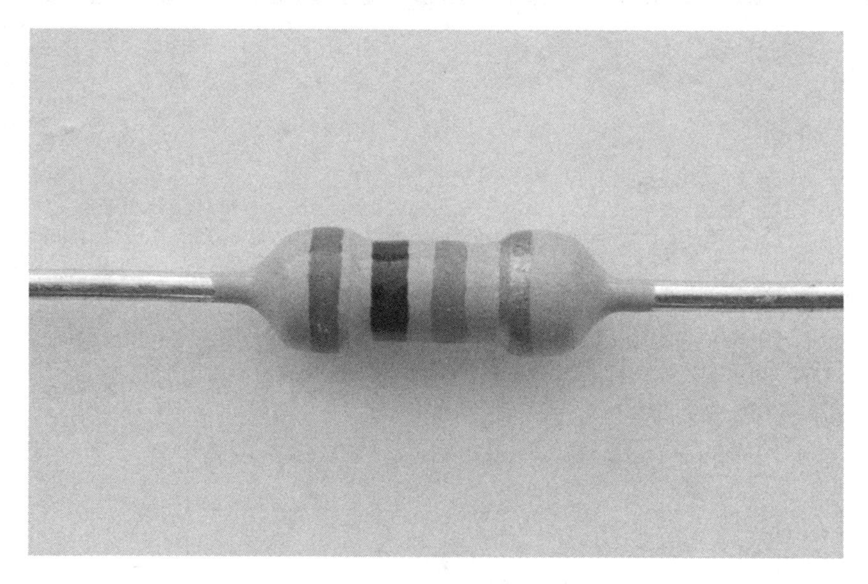

Figure C-1. *This resistor has the color bands brown, black, orange, and gold. This indicates that it's a 10K resistor with a 5% tolerance.*

Table C-2. *10K Ohm resistor with 5% tolerance*

Band	First Band	Second Band	Third Band (Multiplier)	Fourth Band (Tolerance)
Color	Brown	Black	Orange	Gold
Value	1	0	x1K	5%

If you encounter a resistor with five color bands, it simply means that there's an extra digit before the multiplier. See Table C-3 to determine the value.

Table C-3. *Basic five-band resistor reference*

Color	First Band	Second Band	Third Band	Fourth Band (Multiplier)	Fifth Band (Tolerance)
Black	0	0	0	x1	
Brown	1	1	1	x10	1%
Red	2	2	2	x100	2%
Orange	3	3	3	x1K	
Yellow	4	4	4	x10K	
Green	5	5	5	x100K	
Blue	6	6	6	x1M	

Color	First Band	Second Band	Third Band	Fourth Band (Multiplier)	Fifth Band (Tolerance)
Purple	7	7	7	x10M	
Gray	8	8	8		
White	9	9	9		
Gold					5%
Silver					10%

D/Creating a MicroSD Image

Certain features on Galileo are only available if you boot from a microSD card instead of the on-board flash memory. This is because the amount of on-board storage is too limited to include all the possible features. For instance, if you'd like to use Python or Node.js to program the board or if you'd like to use a PCI Express WiFi card, you'll need to download the Intel-provided operating system, load it onto a microSD card, and have your Galileo boot from it.

To do this, you'll need:

- MicroSD card in any size from 1 gigabyte to 32 gigabytes.
- A card reader for writing the data to the MicroSD card from your computer. If you have a standard SD card slot on your computer, you can use an adapter as well.

Here's how to create the SD card and boot off of it:

1. Download the file labeled "LINUX IMAGE FOR SD for Intel Galileo" from Intel's download site (*https://communities.intel.com/community/makers/software/drivers*).
2. If you're running **Mac OS X**:
    a. Download a 7Zip file expander such as Keka (*http://www.kekaosx.com/en/*) or Ex7z (*https://www.macupdate.com/app/mac/19139/ez7z*).
    b. Open the Linux image from Intel with your 7Zip file expander software and expand it to a temporary directory.

c. Insert the microSD card into your computer.

d. Open Disk Utility, found in */Applications/Utilities*.

e. Choose your microSD card from the pane on the left. Make absolutely sure you've selected the right disk. If you're unsure, simply remove and insert the card to see which gets added to the list.

f. In the pane on the right, click the Erase tab.

g. In the Format drop-down menu, choose MS-DOS (FAT).

h. Name it anything you want.

i. Click Erase. This will totally format the card, so you'll lose all the data that may be on it.

j. Find the files you expanded from the Intel image file.

k. Copy all the files together to the root of your newly formatted card. When you open a window to browse the files on your card, you should have a *boot* folder and three other files.

3. If you're running **Windows**:

   a. Download and install 7-Zip from *http://www.7-zip.org/*.

   b. Right-click the *.7z* file downloaded in step 1 and click 7-Zip→Extract Here.

   c. Navigate into the newly created folder, which will be called `LINUX_IMAGE_FOR_SD_Intel_Galileo_v0.7.5` (or similar).

   d. Select all the files, right-click and click Copy.

   e. Insert the microSD card into your computer.

   f. Windows will ask what you'd like to do. Select "Open folder to view files."

   g. If there are any files already on the card, delete them.

   h. Right-click in the drive window and click Paste to copy the files from the folder you extracted. When you browse the files on your card, you should have a *boot* folder and three other files.

4. If you're running **Linux**:

   a. If you don't have it already, install p7zip with the command `sudo apt-get install p7zip`.

   b. Navigate to the folder where you downloaded the *.7z* file.

   c. Execute `7zr x LINUX_IMAGE_FOR_SD_Intel_Galileo_v0.7.5` to extract the file into a folder called `LINUX_IMAGE_FOR_SD_Intel_Galileo_v0.7.5`.

   d. Execute `df -h` and note what drives are available to you.

   e. Insert your MicroSD card.

f. Execute `df -h` to see which card has been added and where it's mounted. (It will likely be something like `/media/UNTITLED`).

g. Delete everything from the card with the command `rm -rf /media/UNTITLED`, replacing `/media/UNTITLED` with the mount point of your device. Be absolutely sure that this is the correct mount point and that you're comfortable with wiping all data from this card.

h. Change to extracted file directory with `cd LINUX_IMAGE_FOR_SD_Intel_Galileo_v0.7.5`

i. Copy all the files in the directory using the command `cp -r * /media/UNTITLED`.

5. Eject or unmount the card from your computer and remove it.
6. With Galileo powered down, insert the card into the microSD slot.
7. Power on the board and it should boot off the card. The first time you boot, it may take some extra time until it's ready to be used.

It's not always immediately apparent that you've done it correctly, but if you watch the LED labeled "SD" next to the MicroSD card slot, it should blink and flicker quite a bit when you boot up your board.

E/Setting Up Galileo on Windows

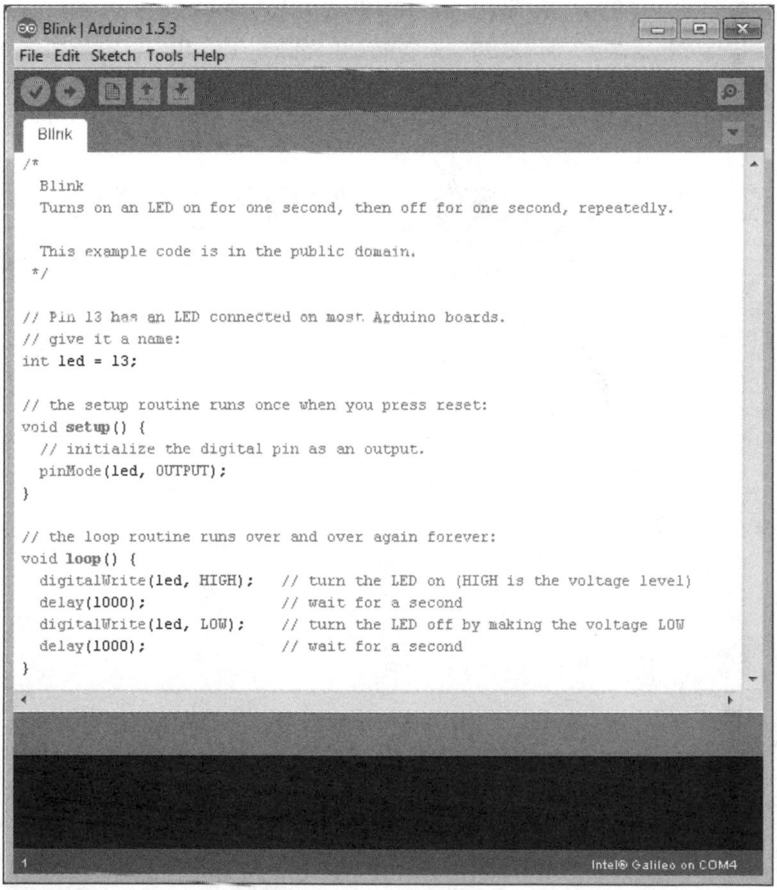

Blink | Arduino 1.5.3

File Edit Sketch Tools Help

Blink

```
/*
  Blink
  Turns on an LED on for one second, then off for one second, repeatedly.

  This example code is in the public domain.
 */

// Pin 13 has an LED connected on most Arduino boards.
// give it a name:
int led = 13;

// the setup routine runs once when you press reset:
void setup() {
  // initialize the digital pin as an output.
  pinMode(led, OUTPUT);
}

// the loop routine runs over and over again forever:
void loop() {
  digitalWrite(led, HIGH);   // turn the LED on (HIGH is the voltage level)
  delay(1000);               // wait for a second
  digitalWrite(led, LOW);    // turn the LED off by making the voltage LOW
  delay(1000);               // wait for a second
}
```

1 Intel® Galileo on COM4

In Chapter 2, you got a quick rundown of how to get up and running with Galileo in Windows.

If you need a little bit more guidance, this section is for you.

The Galileo software is only supported on Windows versions 7 and 8.

1. Navigate to *http://www.intel.com/support/galileo/* and click Software Downloads.
2. Click the link to download "Intel Galileo Arduino SW 1.5.3 on Windows."
3. If your browser asks you, click Open. If not, click on the downloaded file in your browser to open the ZIP file.
4. Drag the entire *arduino-1.5.3* folder onto your `C:` drive. (Figure E-1)

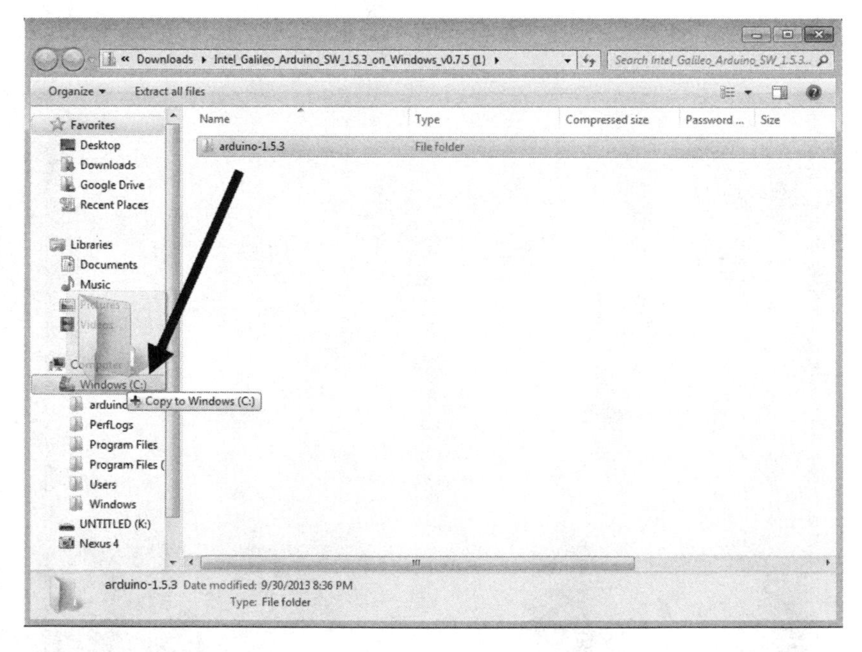

Figure E-1. *Drag the arduino-1.5.3 folder from the ZIP file to the* `C:` *drive.*

You could copy this folder to your *Program Files* folder, but because of an issue with the way files are unzipped in Windows, you'll need to use a different ZIP utility like 7-Zip (*http://www.7-zip.org/*).

5. Open the *arduino-1.5.3* folder within your `C:` drive and double-click on the program `arduino` to launch the Arduino IDE.
6. If you'd like to be able to access the IDE more easily, click on the *arduino* file and click Pin to Start Menu.

After installing the Arduino IDE in Windows, there are a few extra steps you'll need to follow in order to get the IDE to be able to communicate with your board.

1. With your Galileo powered on through the 5-volt power jack, connect it to your computer via USB.
2. You may get an error saying that there was a problem installing drivers (Figure E-2). You can ignore this.

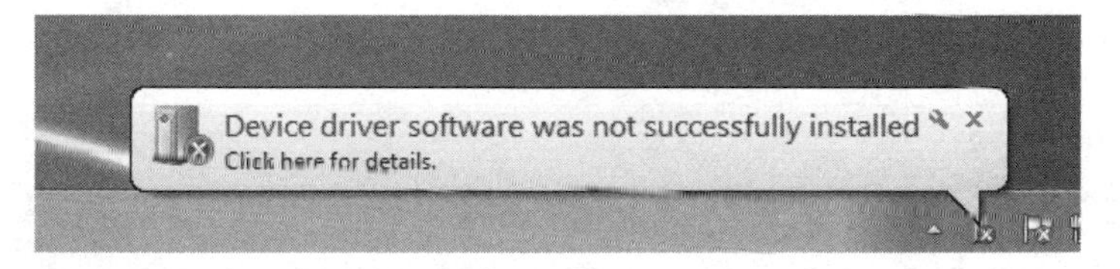

Figure E-2. *When you plug in your Galileo for the first time, you can safely ignore this error message.*

3. Click Start→Control Panel→System and Security→Device Manager.
4. Under Ports (COM & LPT), right-click Gadget Serial V2.4, and select Update Driver Software. (Figure E-3)

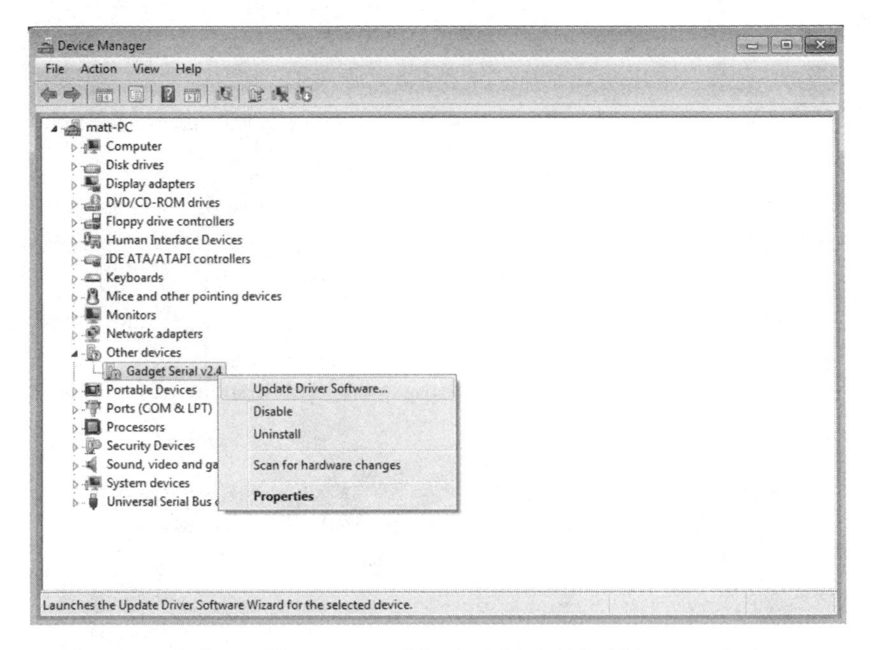

Figure E-3. *Galileo will appear as "Gadget Serial V2.4" in your device manager before you install the driver.*

If you don't see Gadget Serial V2.4 under Ports (COM & LPT), but it appears in Other Devices, check the Galileo Support site (*http://www.intel.com/support/galileo/*) for the resolution to that issue.

5. Click "Browse my computer for driver software."
6. Click Browse... next to the file path input box.
7. If you installed your Arduino IDE folder in the root of your hard drive, navigate to *C:\arduino-1.5.3\hardware\arduino\x86\tools* and click OK. (Figure E-4). Otherwise, you'll have to find where you installed the Arduino folder.

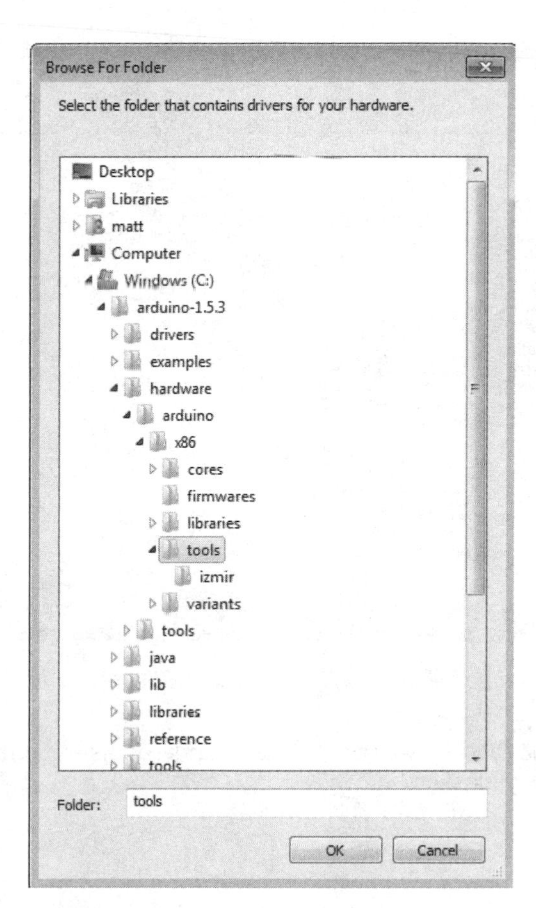

Figure E-4. *Finding the driver in the Arduino folder*

8. You may get a Windows Security alert. Click Install.
9. After the driver has been installed, look under Ports (COM & LPT) in Device Manager again. Note the COM port associated with Galileo (Figure E-5).

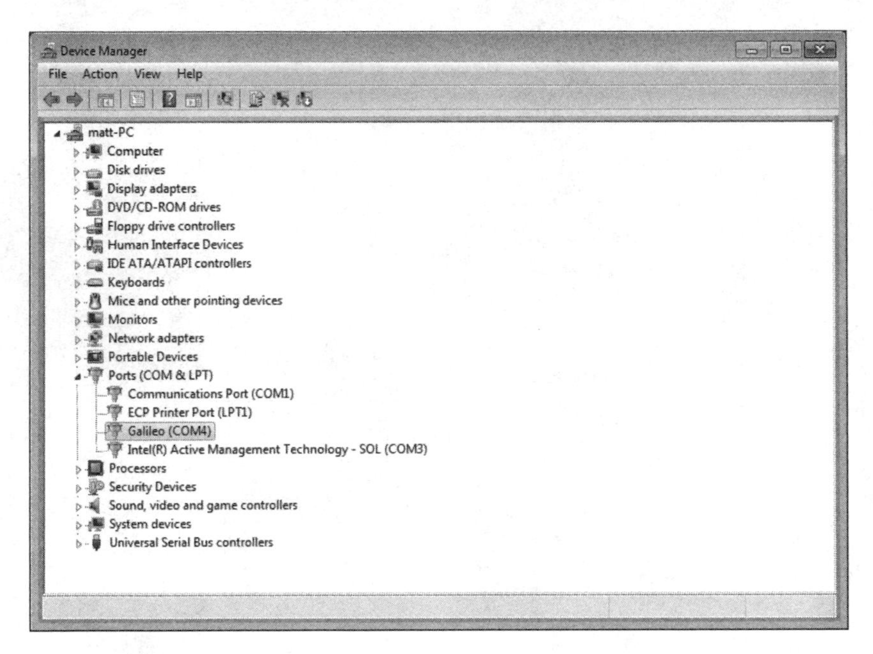

Figure E-5. *Finding the driver in the Arduino folder*

10. Within the IDE, choose this COM port under Tools → Serial Port before uploading your sketch.

F/Setting Up Galileo on Linux

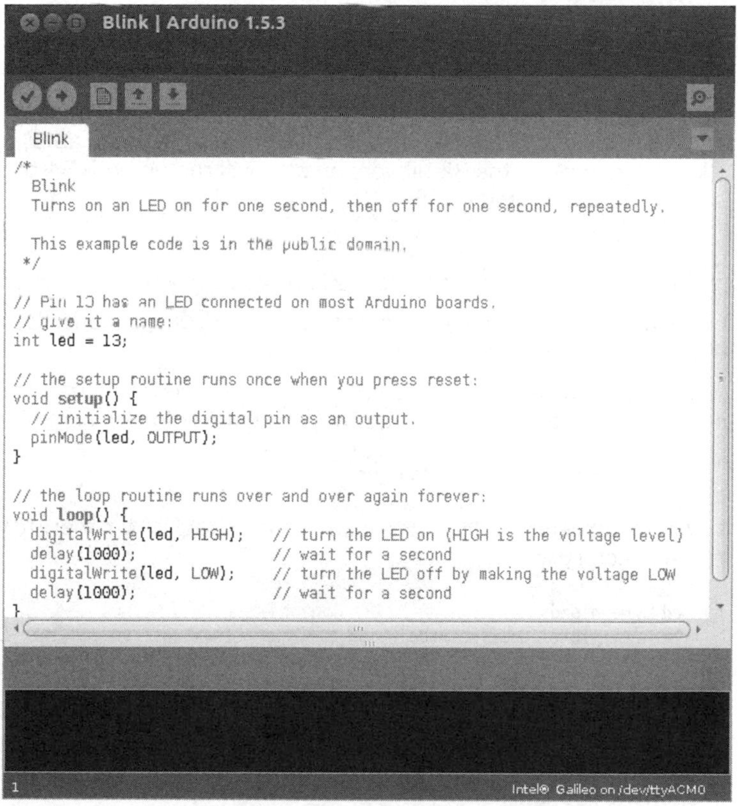

Blink | Arduino 1.5.3

Blink

```
/*
  Blink
  Turns on an LED on for one second, then off for one second, repeatedly.

  This example code is in the public domain.
 */

// Pin 13 has an LED connected on most Arduino boards.
// give it a name:
int led = 13;

// the setup routine runs once when you press reset:
void setup() {
  // initialize the digital pin as an output.
  pinMode(led, OUTPUT);
}

// the loop routine runs over and over again forever:
void loop() {
  digitalWrite(led, HIGH);   // turn the LED on (HIGH is the voltage level)
  delay(1000);               // wait for a second
  digitalWrite(led, LOW);    // turn the LED off by making the voltage LOW
  delay(1000);               // wait for a second
}
```

1 Intel® Galileo on /dev/ttyACM0

In Chapter 2, you got a quick rundown of how to get up and running with Galileo in Linux. If you need a little bit more guidance, this section is for you.

It would be difficult to give exact instructions that work in all Linux distributions. These instructions have been tested on a fresh installation of Ubuntu 12.04.3 LTS. However, they should work on many other distributions as well.

1. Navigate to *http://www.intel.com/support/galileo/* and click Software Downloads.
2. Click the link to download Intel Galileo Arduino SW 1.5.3 on Linux64bit or Intel Galileo Arduino SW 1.5.3 on Linux32bit.

If you're not sure if you're running a 32-bit or 64-bit system, type *uname -m* from the command line. The response will include "32" or "64." If you're still not sure, you can always just download the 32-bit version as it will run on both 64-bit and 32-bit systems.

3. If your browser asks you, save the *.tgz* file in your *~/Downloads* directory.
4. Open a terminal window by clicking on its icon in your application launcher or typing Ctrl+Alt+T.
5. According to the folks at Intel, a system service called modem manager can interfere with the Galileo. Remove it with the following command:

```
sudo apt-get remove modemmanager
```

6. Navigate to where your files downloaded and extract it from there to your home directory with `tar`:

```
cd ~/Downloads
tar -xzf Intel_Galileo_Arduino_SW_1.5.3_on_Linux64bit_v0.7.5.tgz -C ~/
```

7. Navigate to where you extracted the files:

```
cd ~/arduino-1.5.3
```

8. Launch the Arduino IDE:

```
./arduino
```

9. Within the IDE, choose `/dev/ttypACM0` under Tools → Serial Port before uploading your sketch. (Figure F-1)

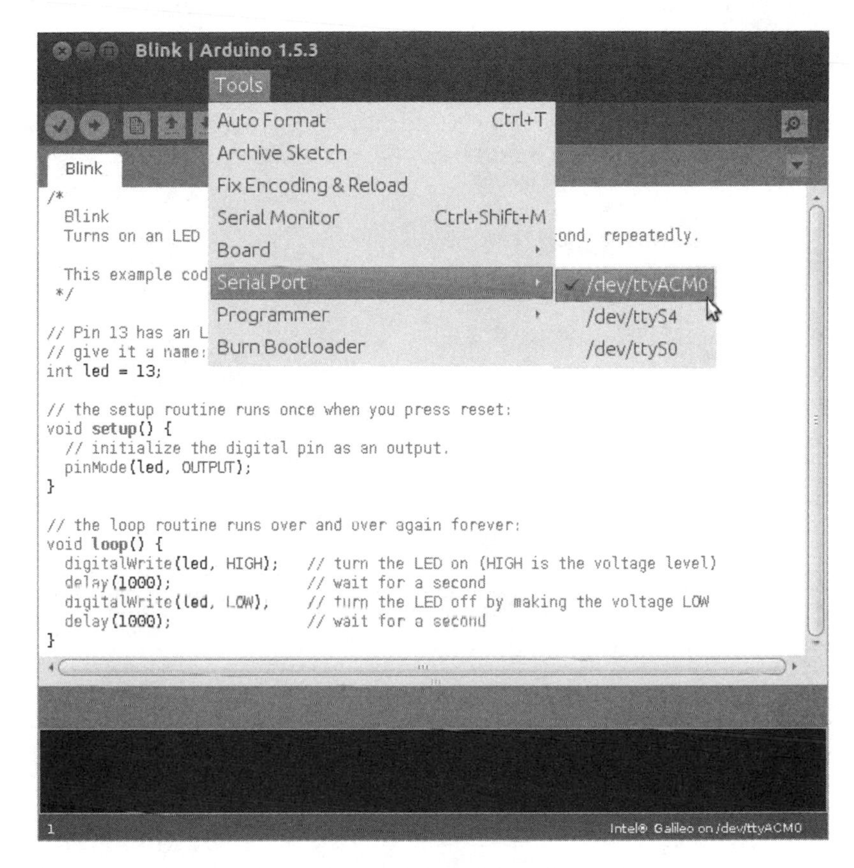

Figure F-1. *Choose the serial port labeled /dev/ttypACM0.*

Linux Notes

If you get an error about Java not being found, install it with:

```
sudo apt-get install default-jre
```

If the serial menu in the Arduino IDE is grayed out, it probably means that you need to escalate to root when launching the Arduino IDE to access the serial port. To do this, exit Arduino and re-execute it as root:

```
sudo ./arduino
```

G/Setting Up Galileo on Mac OS X

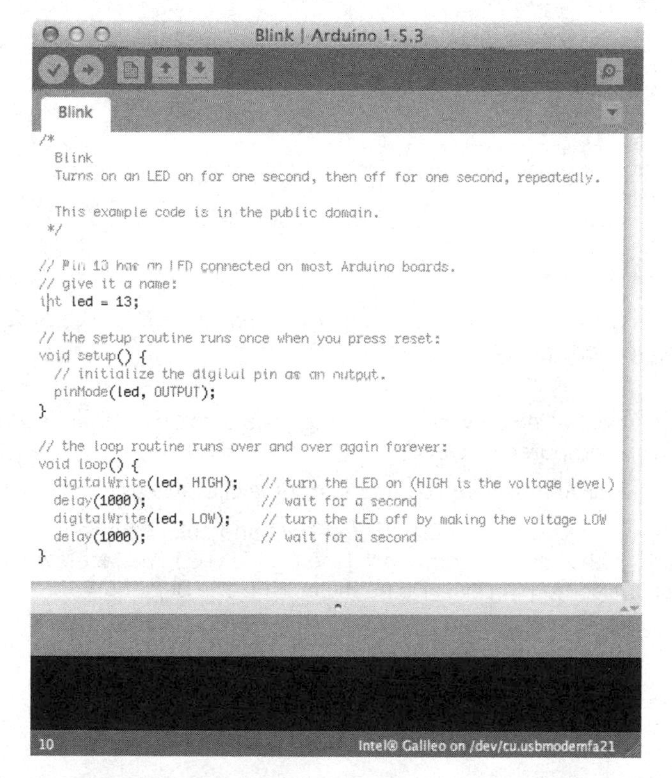

In Chapter 2, you got a quick rundown of how to get up and running with Galileo in Mac OS X. If you need a little bit more guidance, this section is for you.

1. Navigate to *http://www.intel.com/support/galileo/* and click Software Downloads.

2. Click the link to download Intel Galileo Arduino SW 1.5.3 on MacOSX.
3. If your browser asks you, click Open. If not, click on the downloaded file in your browser to open the ZIP file.
4. In your *Downloads* folder, drag the Arduino application that was extracted from the ZIP file into your *Applications* folder (Figure G-1).

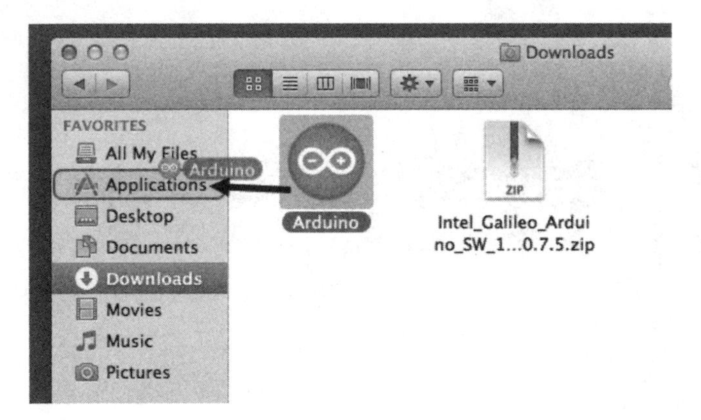

Figure G-1. *Drag the Arduino application from your Downloads folder to your Applications folder.*

5. With your Galileo powered on through the 5-volt power jack, connect it to your computer via USB.
6. Open your *Applications* folder and double click `Arduino` to launch it.
7. Within the IDE, click Tools→Serial Port and select the serial port that starts with */dev/cu.usbmodem* (see Figure G-2) The last few digits will likely be something else. Do not select the one beginning with /dev/tty.

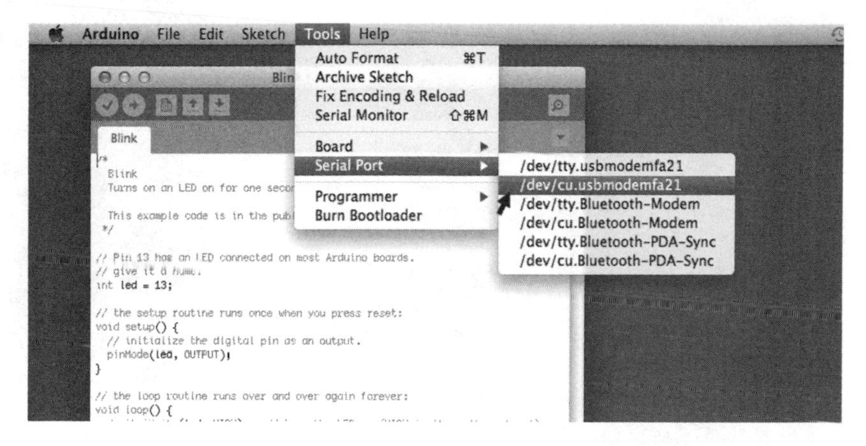

Figure G-2. *Choose the serial port that begins with "/dev/cu.usbmodem"*

Mac OS X Notes

- If you already have a copy of the standard Arduino IDE installed, you can rename the one you downloaded from Intel and keep both in your *Applications* folder. The name you choose must not contain spaces, so just `Galileo` would be a good choice.

H/Connecting to Galileo via Serial

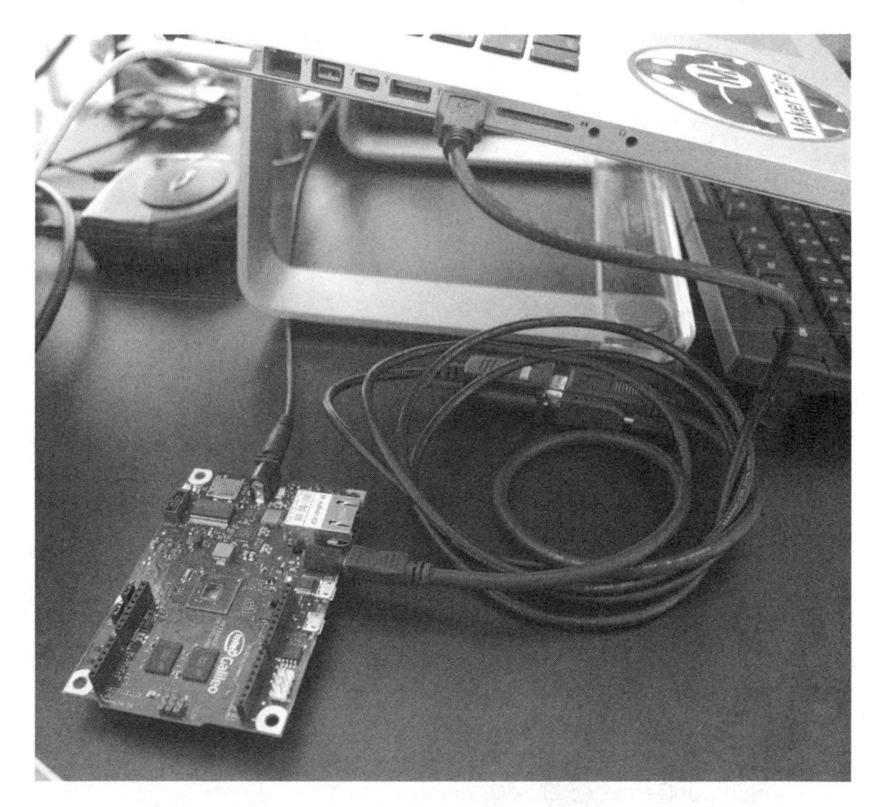

Figure H-1. *Connecting Galileo's serial port to a computer's USB port.*

There are a few different ways to connect to Galileo's command line from your computer. In "Connecting via Telnet" on page 61, you learned how to connect over the network via Telnet. This appendix will show you how to connect to the command line using a serial cable.

This method lets you run Linux commands without any networking, which will come in handy if you're trying to debug a networking problem. It also lets you see the debug messages that Galileo outputs while it's booting up. That information might be helpful in case there's some other problem with the board.

While this connection is made via serial, it's different than the connection that's made in the serial monitor of the Arduino development environment. Arduino's serial monitor is for viewing output from your Arduino code and not for command line use.

To connect to the command line via serial, you'll need:

- DB9 Female to 3.5mm serial cable, such as this one from Amazon (*http://amzn.to/1hSKs9Y*).
- USB to RS-232 DB9 serial cable if your computer doesn't already have a DB9 serial port (and many don't nowadays). This cable from Amazon (*http://amzn.to/1lEZKCP*) will work.
- CoolTerm (*http://freeware.the-meiers.org/*), which is a free, cross-platform terminal utility. You could also use the built-in `screen` command-line utility on Linux or OS X (for example, `screen /dev/tty.usbserial 115200`).

To connect via serial:

1. Make sure your Galileo is powered properly.
2. Connect the DB9 to 3.5mm cable to the DB9 port on your computer or USB adapter cable.
3. Connect the 3.5mm side of the cable to the 3.5mm jack on Galileo.
4. Launch CoolTerm and click Options.
5. If you're using the USB adapter cable:
    a. Open the port listing to check the available serial ports.
    b. Insert the USB cable into your computer (see Figure H-1) and then click Re-Scan Serial Ports.

c. If there's a new port available in the port listing, select that port. If nothing appears, you may have to install drivers for your particular adapter cable.

6. If you're using the DB9 serial port on your computer, you may have to try each of the listed ports to see which works.
7. For Baudrate, select 115200 (Figure H-2).

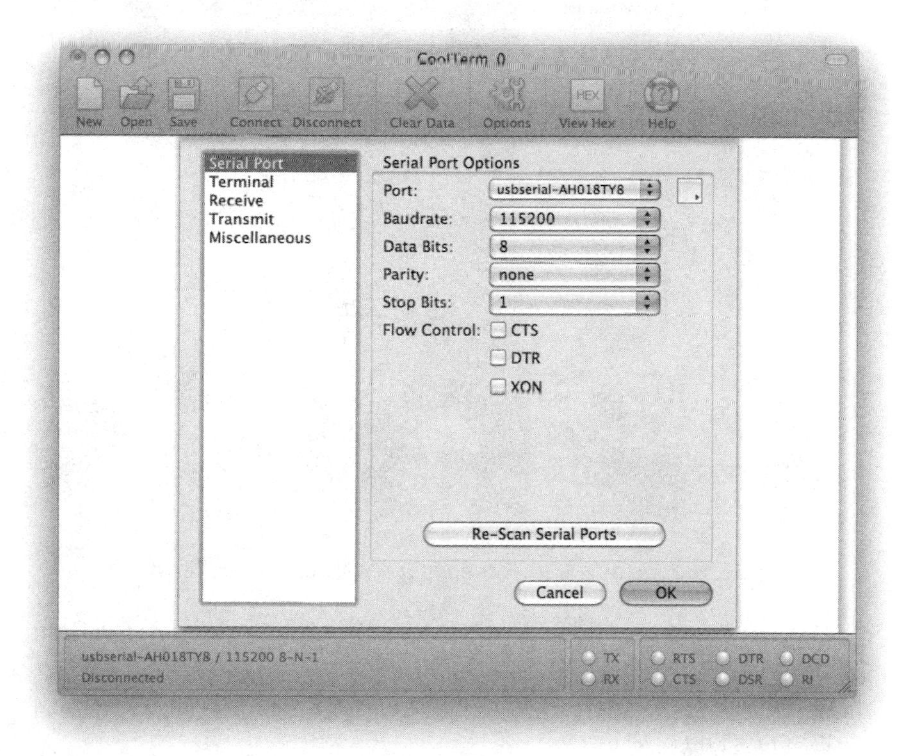

Figure H-2. *Selecting the speed and serial port in CoolTerm.*

8. Click Terminal in the listing on the left to show additional options.
9. Choose CR for Enter Key Emulation.
10. Check the box for "Handle BackSpace Character" to enable it (Figure H-3).

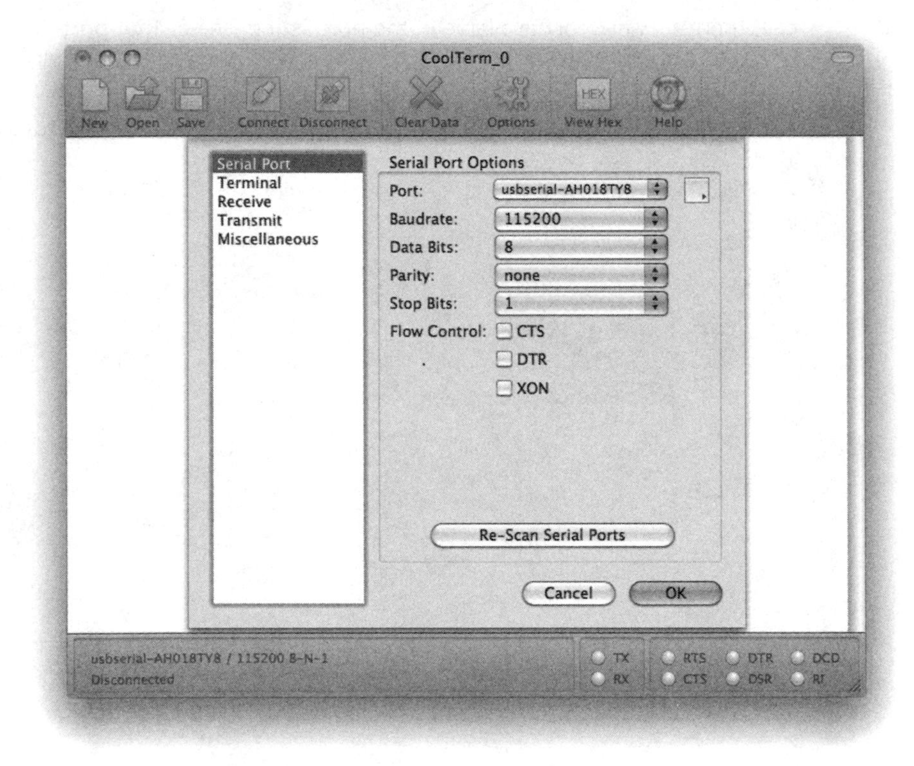

Figure H-3. *Selecting the speed and serial port in CoolTerm*

11. Click OK to accept those settings and go back to the main CoolTerm window.
12. Click Connect in CoolTerm's toolbar.
13. Press Enter.
14. You should see a login prompt appear:

```
Poky 9.0 (Yocto Project 1.4 Reference Distro) 1.4.1 clanton /dev/ttyS1


clanton login:
```

15. Log in with the username `root` and press Enter (Figure H-4).

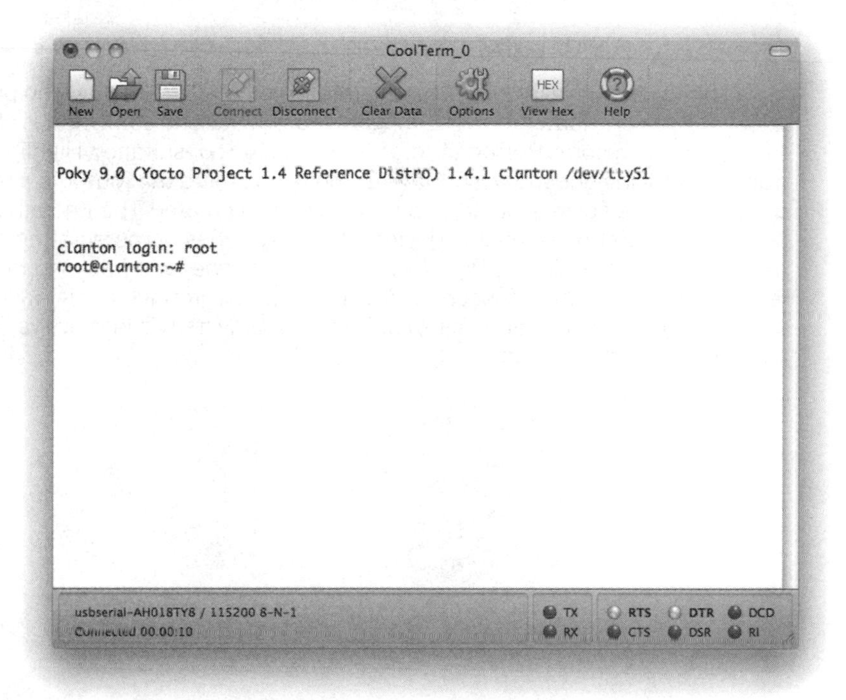

Figure H-4. *CoolTerm connected to Galileo and logged in*

After that process, you should be at your Galileo's command line:

```
root@clanton:~#
```

Remember that any changes you make to files will not persist after a reboot unless you're booting from a microSD card. See Appendix D for how to create one.

About the Author

Matt Richardson is a Brooklyn-based creative technologist and video producer. He's a contributor to *MAKE* magazine and *Makezine.com*. Matt is also the owner of Awesome Button Studios, a technology consultancy. Highlights from his work include the Descriptive Camera, a camera that outputs a text description of a scene instead of a photo. He also created The Enough Already, a DIY celebrity-silencing device. Matt's work has garnered attention from *The New York Times*, *Wired*, *New York Magazine*, and has also been featured at The Nevada Museum of Art and at the Santorini Bienniele. He is currently a resident research fellow at New York University's Interactive Telecommunications Program.

The cover photo is by Jeffrey Braverman. The cover fonts are Benton Sans and Soho. The text font is Benton Sans; the heading font is Serifa; and the code font is The Sans Mono.